Christopher Smith is one of those people whose idea of a good read is browsing through a dictionary. He gets as much fun out of a rare word as an ornithologist does out of sighting an avocet. At the University of East Anglia he teaches French Language and Literature, taking a special interest in drama. What spare time he has he spends singing and reviewing plays, concerts and performances of operas.

Also available from Century Hutchinson
Batty, Bloomers and Boycott **Rosie Boycott**

Alabaster, Bikinis and Calvados

an ABC of toponymous words

Christopher Smith

C

CENTURY PUBLISHING

LONDON

Set in VIP Baskerville by
D. P. Media Ltd., Hitchin, Hertfordshire

First published in Great Britain in 1985 by
Century Hutchinson Publishing Ltd.,
Portland House,
12–13 Greek Street, London W1V 5LE

Printed and bound in Great Britain by
Anchor Brendon Ltd., Tiptree, Essex

British Library Cataloguing in Publication Data

Smith, Christopher
Alabaster, bikinis and calvados: an ABC of toponymous words.
1. Gazetteers 2. Eponyms
I. Title
910′.014 G103.5

ISBN 0-7126-1071-5

Introduction

According to the Book of Genesis, it was God who decreed that the light should be called day and the darkness should be called night. After that, however, He told Adam that it was up to him to name all living things. Quite how Adam set about it, without any hints from Eve to help him, remains one of the many mysteries of the account of creation. But when you come to think of it, it must have been quite a job to decide what was the right label in each particular case. Most people are fortunately in a much easier position with regard to language, and when the need for naming something new arises, the past is generally there to offer a fair amount of help. Etymology is the name for the study of the origins of words, and great fun it can be to peel off layer after layer of change and get back to the beginnings. There is a danger, of course, and it lies in imagining that a word's origin is a reliable guide to its current meaning and, even worse, in attempting to regulate current usage by referring back to the word's roots. Nevertheless, if we can avoid that pitfall, the study of the origins of words can make us savour our language all the more.

The class of words considered in this survey is that of toponyms – that is to say, words founded on place names, the Greek noun *topos* meaning 'a place'. In *Batty, Bloomers and Boycott* Rosie Boycott, for reasons which hardly require speculation, sifted through the rich crop of English eponyms – that is to say, words derived from the names of people. My aim is to do the same thing for toponyms, not hoping to list every single one but making a selection, picking them from different spheres of human activity. Some are obvious once they are pointed out, though they may have been so well assimilated into our everyday speech that we hardly notice them; others are buried deep beneath centuries of linguistic and cultural change. Nearly all of them have come into English with a rich freight of associations and history. An island off the western coast of Europe, Britain has traded with the Old World and the New, and commercial links have brought words into our vocabulary as well as gold into our coffers.

There are more toponyms than you might expect and they have come from many different places – from Shanghai in China to Lydd in Kent. Go to the greengrocer's and you can find tangerines (from Tangiers), satsumas (from a Japanese province) and peaches (from Persia, originally); and the vegetables you purchase may well include swedes and Brussels sprouts. The soldier fixing his bayonet might like to think of the fortress city of Bayonne, and in the Second World War he would ride into battle in a Bren-gun carrier, which had emerged into the language by a circuitous route. Fashion and fortification, food and drink, architecture and sport and a whole range of activities besides use toponyms in their vocabulary, reminding us that for centuries Britain has stood at the cross-roads of the world. Many, as you might expect, are dim reflections of export trades of long ago, while others commemorate episodes from history or, less exciting but still part of our heritage, remind us where our forefathers went for rest and recreation. Some toponyms remain easily recognizable, but there are others which lie hidden and need a few words of explanation. An argosy is the richer when you know what port it sailed from, and before you dismiss this as bunkum, you should know where that all started.

A number of toponyms are connected with the needs of the consumer. Many of the factors governing the acceptability of a word in our language will be on display when you make a visit to the local market, for example. There is a constant search for terms which will indicate precisely what it is you are buying, but which will not sound too grand or too foreign, or simply take too long to read or say. Over the years one easy way of inventing new terms has been to use the name of the place of origin in some form or another. That can give an aura of fashion in some cases and, provided the new word does not prove too difficult to pronounce, it is likely to stick in the language. In the first instance it may be a matter of using an adjective from a foreign place name to qualify an English noun, but after a while the sense will become clear enough and most people will get by with using the adjective as a noun in its own right. The variations and permutations are endless, and the only safe generalization about language seems to be that rules never apply in the same way in every case. For instance, there is the puzzle as to why tuxedo – which everybody in Britain understands perfectly well – has never caught on, leaving us with the dreary, unwieldy and not even particularly accurate name, dinner jacket.

A list of toponyms could be expanded *ad lib* from China to Peru, by including all those words referring to places commonly used in English simply to indicate where products come from. But it would cloud the issue to list Barsac and Beaune, Chevenette and Chinon and so on until we come to the last page of André Simon's *Gazetteer of Wines*, when every toper knows that the place name simply tells us where the grapes are at least supposed to have been grown. It seems better to pick and choose among toponyms, selecting those which have a story to them or which illustrate interestingly some aspect of the evolution of our language.

What's in a name? It is hard to say precisely, but that toponyms are still reckoned to have power in them is demonstrated by the care taken to find the name Lymeswold for the newly developed English cheese. Why did the ad-men reject every other name in the gazetteer? And what message were they sending by their choice?

Toponyms are not being created at any great rate in the present day apparently, and it is worth pausing to wonder why this should be. There appears to have been a time when – perhaps because the transport of raw materials was reckoned prohibitively expensive and because trade guilds ensured that the tricks of the various trades were kept pretty confidential – localities were able to maintain something of a monopoly on the production of various commodities. Cloth manufacture would be a good case in point. Geographical discoveries have had their impact on toponyms too. In the ancient world there were obviously fewer places where, for instance, a mineral might be found than there are nowadays when prospectors have travelled over the whole world and laboratories are offering synthetic substances. Centres of high culture or outstandingly advanced technology exported goods (and words for them too) to less favoured countries all around, and even if a rival product was eventually brought out, it often borrowed the imported name in unconscious tribute to what it was imitating. Nowadays, in an age of international trade and multinational technology when many speakers are offended by anything smacking of linguistic imperialism, the climate is far less favourable to toponyms. Perhaps that is just one more reason for enjoying those which still enrich our vocabulary.

In addition to existing as nouns and adjectives

which we can use in the normal ways, toponyms have their place in a number of idioms. 'All shipshape and Bristol fashion' is an old expression which has survived the competition it had to face in the 1890s from 'All Sir Garnett', a reference to the reforming zeal of Sir Garnett Wolsey which so impressed the Victorians. 'Castles in Spain' still attract some, though the idiom is gradually disappearing, perhaps because nowadays the Costa Brava means something more real to most people. But the odious practice of 'sending people to Coventry' continues, even if there is still some doubt as to how the expression came to be coined.

In preparing this little book I have turned to a number of specialized works on various trades, sciences and so on. Naturally I have also consulted the standard reference works in the field; among them are Eric Partridge's magisterial *Name into Word*, Webster's *Third New International Dictionary*, *The Longman Dictionary of the English Language*, the fine French dictionaries of Emile Littré and Paul Robert and, of course, the *Oxford English Dictionary* (which figures here from time to time under the initials *O.E.D.*).

I must thank Ann Cook for typing like a Trojan and my wife and daughter for suggestions, corrections and cheerful help.

Aberdeen-Angus
It is rather satisfactory to begin with a Siamese-twin toponym. There are ample records of black hornless cattle in Aberdeenshire and Angus, and in 1808 the breed as we know it was founded when one Hugh Watson of Keillor, near Coupar Angus, purchased the blackest and best hornless heifers available at Trinity Fair, near Brechin in Angus, and crossed them with local black-polled bulls. The next stage came when William McCombie of Tillyfour, in Aberdeenshire, improved the breed by crossing cows from Aberdeenshire with bulls brought in from Angus. He proved the superiority of the resulting cattle by winning championships at Smithfield in 1867 and at the Paris Exhibition in 1878. Now the breed is known as far away as New Zealand and Argentina. Black and hardy, capable of withstanding cold winters and hot summers, the Aberdeen-Angus is one of the most famous beef breeds.

Acadialite
Acadialite is a reddish coloured chabazite, that is to say a glassy mineral composed of silica, alumina and lime. It is so called because it is found in Nova Scotia, which the French called Acadia from the time of the first attempts at colonization there by French settlers in 1604 until it was definitively ceded to England under the Treaty of Utrecht in 1713. The name lives on in one of Nova Scotia's institutes of higher education, the University of Acadia at Wolfville.

Alabaster
Alabaster, the translucent and sometimes attractively veined gypsum which is suitable for carving ornaments and monuments and takes a high

polish, refers – according to Pliny – to Albastron, a city in Ancient Egypt. The best alabaster is said to come from Volterra in Tuscany, however.

Alka-Seltzer

Alka-Seltzer is Bayer's trade-name for an effervescent antacid pain reliever which has been blessed by many on the morning after. Compounded of aspirin, citric acid and sodium bicarbonate, Alka-Seltzer looks back for its name to the Seltzer water which was originally obtained from the springs in the German village of Nieder-Selters, in Hesse, a dozen miles south-east of Limburg. The naturally effervescent water contains sodium chloride and small quantities of carbonate of sodium, calcium and magnesium. In Edwardian times, hock and seltzer was a popular refreshing 'long' drink.

Alsatians

These powerful intelligent dogs were originally known in this country – by a simple translation of their native name – as German Shepherd Dogs. But in 1917 when things German were generally taboo and even English aristocrats were thinking it wise to change or adapt their names, the form 'Alsatian' was born. It is true that there was no essential connection between the dogs and Alsace, but the idea of associating them with the province which Germany had wrested from France in 1871, and whose recovery was high amongst war aims, had obvious attractions. Nowadays dog-lovers are reverting to the earlier usage, no doubt partly out of a desire for correctness, but probably also because police and security work have given Alsatians a bad name.

Angostura Bitters

This is one of the most entertaining toponyms because the place has changed its name while the product named after it has remained unaltered. Dr Siegert was an army surgeon of Silesian origin who, after service in the wars against Napoleon, crossed the Atlantic and joined Simon Bolivar in Venezuela. To help the soldiery over stomach and digestive disorders he developed a recipe for aromatic bitters and by 1830 had begun to export them to Trinidad and then on to England. At first they were known as Dr Siegert's Bitters, but because they were manufactured in the town of Angostura, on the River Orinoco 240 miles from the sea, they came to be called Angostura Bitters. In 1849 Angostura was renamed Ciudad Bolivar, in honour of the South American liberator; by then, however, the name of the drink was firmly fixed.

The exact composition of Angostura Bitters is kept secret and the microscopic print on the bottle label is enough to deter all but the most curious. Still, the great usefulness of Angostura Bitters is no mystery. The Royal Navy has not only traditionally watered rum to make grog, but has used Angostura Bitters to make pink gin. A couple of drops of Angostura Bitters are swirled round the glass, then a measure of gin and another of iced water added. Ideally it should be Plymouth gin – another toponym, of course – which is made by Coates & Co. of Plymouth; in aromatic quality it comes mid-way between London gin and Hollands (or Dutch) gin.

Angora cats

With their long silky white coats, Angora cats are said to have been introduced into Europe by travellers returning from the capital of Turkey, now known under the spelling Ankara. The breed suffered something of an eclipse when interest focused more on pure Persian cats, but after new imports from Turkey the Angora has been making something of a come-back.

Antimacassar

We may need a special term for antimacassars, which look like a denial of a potentially bogus toponym. In theory they were used to prevent the tall backs of armchairs from being soiled by the macassar oil with which Victorian gentlemen anointed their hair. In fact they became excuses for ornamentation and embroidery, laundering and starching and all that conspicuous expenditure of women's time, effort and eyesight which the Victorians pretended was a sign of domestic propriety. Whether macassar oil really came from Macassar, a district in Celebes in the East Indies, is also a matter of some dispute. The original vendors of the hair oil, Rowland & Son, ran a great advertising campaign in the early nineteenth century in which they assured the public that this was so. But according to the late nineteenth-century analyst quoted in *O.E.D.*, 'what at present comes into commerce under the name of "macassar oil" is mostly a mixture of cocoa-nut oil and ylang-ylang extract coloured red with alkanin', and the 1911 edition of *Encyclopaedia Britannica* states baldly that 'Macassar oil is a trade name, not geographical'.

Arabic numerals

The Renaissance saw the triumph of roman and italic letter forms – derived respectively from ancient Roman inscriptions and the manuscript hand of the humanist scholars – over the unwieldy forms of the Gothic alphabets. But because they were clumsy and, furthermore, lacked a symbol for nought, roman numerals had begun to be displaced by arabic figures at an earlier date. The two systems exist side by side to the present day, but of the advantages of the arabic numerals, which probably derive from Indian developments in the eighth century AD, there can be no doubt.

Aran patterns

The three Aran islands are off the west coast of Ireland, forming as it were a natural breakwater across the opening of the great inlet of Galway Bay. They are famous for the number of architectural remains dating from the first century AD onwards, including many related to early Christianity. John Millington Synge spent a number of summers on the islands at the turn of the century; he evoked the poverty and the poetry of the place in *The Aran Islands* and also in his powerful short drama *The Riders to the Sea*, first performed in 1904 and subsequently turned into an opera by Ralph Vaughan Williams. Aran sweaters made of undyed oiled wool became high fashion in the seventies; the knitting of their traditional richly textured ornamental patterns characteristically involves the use of a third needle to make cables and projecting 'blackberries'. Another island famous for its knitting is of course Fair Isle, midway between Orkney and Shetland. The brightly coloured Fair Isle patterns became very popular between the Wars, particularly after the Prince of Wales and the Duke of York took to wearing Fair Isle pullovers with plus-fours for golf.

Argosy

Argosy, a vessel with a rich cargo and possibly even laden with dreams, sailed into English usage in the sixteenth century from Ragusa, the seaport on the fabulous Dalmatian coast which has been known as Dubrovnik since it became part of Yugoslavia.

Artesian wells

The great thing about artesian wells is that you don't have to lower a bucket and then winch it up. The idea is to make a narrow bore-hole down to the lower levels of a water-bearing stratum shaped like a saucer; the pressure of the water in the higher parts of the saucer then forces it up to the surface through the bore-hole. At least that is the theory, but bore-holes fitted with pumps are also commonly called artesian wells. The name comes from the French province of Artois where wells of this kind were often bored, though in ancient times the technique had been employed in many parts of the world, in Asia Minor, Persia, Egypt and China.

Arras

Poor Polonius, dead for a ducat, paid the price of hiding behind an arras while eavesdropping on Hamlet and his mother. An arras became a general name for rich and heavy wall hangings because the city of Arras, in northern France in the department of Pas-de-Calais, was famed in the late Middle Ages as a centre for the weaving of fine tapestries.

Astrakhan

The soft, thickly curled wool which provides Soviet statesmen with hats and collars to keep out the cold during even the longest ceremonies at Lenin's Mausoleum comes, at least in theory, from very young karakul lambs from Astrakhan, which lies where the River Volga flows into the Caspian Sea.

Attic

As well as Attic salt, the elegant wit with which Athenians were supposed to give savour to their conversation, Attica has contributed the architectural term 'attic'. First applied to the masonry going up above the cornice of a building, it is now used loosely to mean rather low rooms at the top of a house and sometimes built into the roof itself.

Ayrshires

Early in the history of the breed, Ayrshires might have posed some tricky questions of categorization as either eponyms or toponyms. Bred by one John Dunlop, of Dunlop House in the parish of Ayrshire in Scotland, they were originally called Dunlop cattle – though whether after the man, his home or its location is a moot point. Next they became Cunninghams, after the district where Dunlop is situated, and it was only subsequently that they became Ayrshires. John Dunlop presented Robert Burns with one of his heifers and the poet was delighted with the gift. Quite how the breed was developed is uncertain, but it seems clear enough that the poor early eighteenth-century cattle of the region were crossed with Teeswater, Jersey and perhaps even Dutch cattle. Originally black and white, Ayrshires are now brown and white. They are held in high regard for their ability to produce large quantities of rich milk in relatively severe conditions.

Badminton

Badminton is the name of the country seat of the Dukes of Beaufort in Gloucestershire. The game played with rackets and shuttles is said to have evolved out of battledore and shuttlecock around 1870 at Badminton. Taken out to India by army officers, the rules of badminton were first drawn up at Poona a few years later. Since then it has spread worldwide, another testimony to Britain's ability to devise games even if we cannot contrive to win them very often these days. The Badminton Library is the title of a famous series of books on sport and games; Badminton is also the name of a drink made by diluting claret with soda water.

Balaclava helmet

There is a rich irony in the name Balaclava helmet as a reminder of the Crimean War, the nature of which was not inaptly summed up by its opponents in the simple anagram of 'a crime'. The Battle of Balaclava, fought on 25 October 1854, is famous for the futile valour of the Light Brigade charging under the witless leadership of Lord Cardigan, though if we were fair (and read just a little more of our Tennyson) we might also recall that the Heavy Brigade had scored a brilliant success on the morning of the same autumn day. But whether the helmet is called after the battle is doubtful; it seems more likely that the name commemorates the long cold winter during which troops sat and shivered around the supply base of the British army at Balaclava itself, down by the sea. The term Balaclava helmet is recorded as being first used in 1892, some forty years later than the campaign after which it is named. Given the slowness with which the War Office responded to the needs of the troops in the Crimea, it might be imagined that somebody had at last got around to dealing with the requisition from Lord Raglan.

Barbary apes

'The establishment of the apes on Gibraltar should be twenty-four, and every effort should be made to reach this number as soon as possible, and maintain it thereafter.' This was Winston Churchill's memorandum to the Colonial Secretary on 1 September 1944. Even in the midst of war – indeed, perhaps precisely because we were in the midst of war and morale mattered – he recognized the importance of the myth that while the Barbary apes were on the Rock, Gibraltar would remain British. Also he had sense enough to discount the tale that they came across the Straits by way of an under-sea tunnel. The apes come from what used to be called the Barbary Coast – the northern seaboard of Africa from Morocco in the west to Tripoli in the east – 'Barbary' being a corruption of Berber. The tail-less Barbary apes are found in Algeria and Morocco.

Barrack

The verb to barrack – i.e. to shout hearty abuse in stentorian tones (after Stentor, the Greek herald with an exceptionally loud voice at the Trojan War) – at players and umpires or referees at a sporting contest is, as might have been guessed, Australian in origin. Ian Partridge quotes a correspondent who assured him that the term was first applied to the rough-and-ready teams which played football near the Victoria Barracks, Melbourne, in the last two decades of the nineteenth century.

Bath chairs

Invalids taking a cure at Bath had reason to be grateful to James Heath, who invented the Bath chair which was pushed by hand and mounted a folding hood. While in the town, they could of course enjoy Bath buns and Bath Oliver biscuits.

Bauxite

Bauxite is important as one of the raw products from which aluminium is obtained. It is a reddish rock named after the village of Les Baux (or Beaux) – near Arles in the department of Bouches-du-Rhône, in southern France – where it was first mined in significant quantities. Though originally called 'beauxite', the modern form of the name dates back as early as 1861.

Bayonet

Bayonets were originally flat daggers; the suffix *et* suggests they were small ones. They were made at Bayonne,the great fortress town in south-west France on the frontier with Spain. In the mid-seventeenth century someone had the idea of pushing bayonets into musket barrels so that the musketeers were not defenceless against enemy cavalry once they had fired. The next stage was the invention of a clip for fixing the bayonet to the end of the barrel, so that the musketeer could fire another shot once he had had the chance to complete the lengthy process of reloading. The development of the bayonet meant that special corps of pikemen were no longer needed in armies.

Bedlam

'It's bedlam in there.' We all know what that means, though not everyone will be clear whether 'bedlam' is a noun or an adjective, and it has lost the initial capital letter which might have served as a pointer. In fact it refers back to the Hospital of St Mary of Bethlehem which was founded in 1247 by Simon FitzMary, sheriff of London, as a priory for sisters and brothers of the Order of the Star of Bethlehem. Originally sited in Bishopsgate, London, from the fourteenth century onwards it was used for the confinement of the violently insane. With the dissolution of the monasteries, it then passed under the control of the City of London and arrangements were confirmed by Henry VIII in 1547. Though the building survived the Great Fire of London, the premises were judged so inadequate that in the 1670s the institution removed to Moorfields. Gawping at the lunatics was something of a fashionable entertainment in the eighteenth century, and the sights in what by then was called Bedlam inspired a plate in Hogarth's 'Rake's Progress' and, in turn, one of David Hockney's sets for Stravinsky's opera of the same name. In 1815 the Bethlehem Royal Hospital left Moorfields and occupied buildings built to designs by James Lewis in St George's Fields, Lambeth Road, London. There it remained for more than a century, moving out to Eden Park, near Beckenham in Kent, in 1930. The Lewis building subsequently became the home of the Imperial War Museum, where a substantial rebuilding programme is currently taking place.

The name Bethlehem itself means 'House of Bread' and the same root has also yielded Bethel ('House of God') and Bethesda ('House of Mercy'), both of which have given nonconformity apt words for their chapels.

Benitoite

Mineralogists seem to divide into two categories: those who name their discoveries after themselves, thus producing a rich crop of eponyms, and their self-effacing confrères who call them after the place where they found something new. Professor G. D. Louderback is one of the modest ones, which is after all some justification for mentioning him here. In 1907 he described a titano-silicate of barium which is sometimes colourless and sometimes blue, the blue variety being of such a nature that it may be cut as a gem. A brilliant stone with a high refractive index, it appears pale when viewed along its main axis and dark when viewed across. The Professor had discovered the mineral near the headwaters of the river San Benito in California, hence its name.

Some idea of the difficulty of naming discoveries is given by the history of Bentonite, a clay found in the Fort Benton strata of the Cretaceous of Wyoming. In 1898 W. C. Knight noted that he had originally proposed to call it Taylorite, but found the name had already been claimed for another purpose. So here we have a toponym only by second choice.

Benzene

Benzene comes into the category of unguessable toponyms, but beneath the centuries of linguistic development there is a history that adds romance to chemistry. Benzene (or in the first place, Benzine) is the name given in 1835 to the colourless hydrocarbon obtained from coal-tar oil, which was sometimes also called Benzol. This has spawned a whole series of names for organic compounds. The root is taken from benzoin, a word meaning a resinous gum which has been in use since the mid-sixteenth century; it comes from the Arabic *luban jawi*, translated as 'frankincense of Jawa'. An additional complication is that Jawa does not mean Java, as might be surmised, but Sumatra. Transforming *luban jawi* into benzoin involved the dropping of the first syllable. This probably occurred because it was incorrectly assumed to be a definite article, as in such words as algebra. In some instances words from Arabic have been transferred into two forms, with and without article, as in alchemy and chemistry; sometimes the definite article has become permanently attached to the noun which has developed – as happened for instance with the word lute, which was just *ud* or *ut* originally.

Berlin

A berlin was a four-wheeled carriage which took its name from the German city. It was designed around 1670 by an Italian architect in the service of Frederick-William of Brandenburg, the so-called Great Elector. The use of the word in English has not been recorded before 1731. In many countries a 'berlin' is used to mean what we call a saloon car. Probably the term did not catch on in Britain because things German were not particularly popular at the time when cars were becoming more common.

Berlin wool-work

Berlin wool-work was, so to speak, the staple of British embroidery throughout the first half of Queen Victoria's reign. A panel at the Victoria and Albert Museum gives some idea of the style and its social context, depicting in considerable detail a remarkably self-satisfied spaniel elegantly lounging on an oblong cushion with patterned sides and an opulent tassel at each corner. Berlin wool-work was also used for household hangings of every description and such essentials of the nineteenth-century drawing-room as chair-backs and fire-screens. The fashion did originate in Berlin and for some years the richly coloured wools and the squared-paper designs were imported. Eventually local manufacturers saw their chance, but for a time there were shops known specifically as Berlin warehouses. The attractions of Berlin wool-work were its relative ease – for it required no more skill than the patience to follow simple patterns and only relatively simple stitches – and the speed with which quite showy effects could be produced. Protests that Berlin wool-work was debasing the art of embroidery were first voiced in Roman Catholic and High Church circles, with the Gothic revivalist A. W. N. Pugin arguing for a return to medieval styles. In domestic embroidery, however, reform had to wait until William Morris was able to bring about a total change in attitudes.

Bermuda

The Bermuda islands still continue to be fertile in toponyms. Bermuda cigars do not seem to be in great demand these days, but the Bermuda (or Bermudian) rig has conquered the world of racing yachts. Triangular sails with a boom at the foot, but not needing a gaff at the head, were seen in the West Indies as early as 1800, yet it was more than a hundred years before they began to be used for small pleasure boats in Britain. In the twenties they were adopted for the largest racing craft and have not yet been displaced. Things became even more toponymous when – as well as setting a Genoa jib or even, in the thirties, a double-clewed quadrilateral Greta Garbo jib – the great J class yachts set their Bermuda mainsails on Park Avenue booms, which were triangular in section with a wide flat top fitted with rails along which the foot of the sail could run and assume a perfect aerodynamic shape. Bermuda has also given us long shorts which, since they hide the thighs, seem to reveal something about the wearer's character, and the Bermuda triangle, which may be a place but seems likely to blossom into an English idiom for unexplained disasters.

Bikini

Bikinis exploded on French beaches in 1947. The scanty bra and brief pants, originally designed by the Parisian fashion house of Heim, were intended to allow the wearer maximum exposure to the radiation of the sun and the admiring glances of any males fortunate enough to be in the immediate vicinity. Perhaps the idea was not entirely original, for a mosaic from the fourth century AD which has been uncovered – if that's the word – in central Italy shows ladies in similar attire. It might also be thought that no costume could have been less suitable for wearing on Bikini atoll in the Marshall Islands in the Pacific Ocean where the US Government test-fired first atomic bombs and then hydrogen bombs in a series of nuclear experiments between 1946 and 1958. With the passage of time the pants became smaller, and enterprising girls went topless and created the monokini.

Bilbao

A Bilbao – or sometimes a Bilboa – is a type of wall mirror which was fashionable in the seaport towns of New England in the last twenty years of the eighteenth century. The glass was set in a framework of coloured marble or of marble and carved wood. On stylistic grounds it might be supposed that these were the work of followers of the great English furniture makers of the period, but they are believed to have come from the port of Bilbao on Spain's Biscay coast.

Billingsgate

Billingsgate used to be a common euphemism for abusive foul language such as was heard, it was said, at Billingsgate Fishmarket in London. Nowadays it would be calumny to pretend that fishwives swear any more than the rest of us, so the euphemism has gone even if the habit remains.

Blarney

In Blarney Castle, about five miles north-west of Cork, there is a stone which, if it is kissed, confers a gift for persuasive cajolery and easy flattery which has served many an Irishman well. The castle dates from the mid-fifteenth century, but no use of the word blarney is recorded before 1819 – which is odd, because it is hardly the sort of thing people would keep quite about.

Bordeaux mixture

As well as claret, Bordeaux has given us Bordeaux mixture, one of the earliest fungicides still in everyday use. The link is obvious enough. Made like most old-style fungicides with copper in the form of copper sulphate, which is mixed with quicklime and water, it is non-specific in effect and controls a large number of leaf diseases. The drawback is that the lime content can cause stunting to young shoots and damage delicate foliage, especially after light rain. Burgundy mixture is similar, but with washing soda substituted for the quicklime; it too is effective, but can damage delicate plants.

Borstal

Borstal is the name of a village near Rochester in Kent where in 1901 a special reformatory was opened for young adult offenders. As more institutions were set up on similar lines, penologists began to talk about the 'Borstal system'. The sad thing is that 'Borstal boy' has not come to mean 'reformed criminal' in most instances.

Boston

Boston, a card game for four players which is derived from whist and quadrille, was particularly favoured in French high society in the first half of the nineteenth century. It has fallen quite out of fashion, but has left a legacy in the term 'misère', meaning an undertaking to make no tricks. The game goes back to the American War of Independence; Boston has some place in everybody's understanding of that period, if only because of the famous Tea Party. As one of the major ports on the American coast, Boston became the centre of important military and naval operations and the French, for reasons that were not perhaps entirely disinterested, joined in to assist the colonists in the struggle against Great Britain. The game of Boston is said to have been devised by French naval officers serving with the fleet which came to lend support when the town was being besieged by the British; the terms *grande misère* and *petite misère* refer to islands in the vicinity. The full name 'Boston de Versailles' may be connected with the fact that in 1783 a preliminary peace treaty between America and Great Britain was signed in that city, in the house now occupied by the municipal library, and the variant form 'Boston de Fontainebleau' was no doubt simply called after another great French royal city. The popularity of Boston in France was linked with a certain passing fashion for anything connected with America.

Boston rocker

The rocking-chair out on the porch is an important ingredient in the mythology of American decency, and Boston rockers are the most popular of American rocking-chairs. The design is a development of the Windsor rocker, having curved arms, a tall back and a seat which is curved up at the back and down at the front. The top rail is broad, which allows space for stencilling on some suitable design.

Bourbon

The link between Bourbon, pronounced 'burb'n', and the dynasty that ruled in Europe for many centuries is not very strong. In the state of Kentucky, there is a county which received the name of Bourbon to commemorate the aid given to the American colonists by the French during the War of Independence. It was there, in Georgetown, that the Reverend Elijah Craig set up a distillery beside a limestone stream to distil whisky from a mash of what Americans call corn and we call maize. Modern American legislation limits the use of the name Bourbon to whisky distilled in the United States from a mash that is composed at least 51 per cent corn (and usually the proportion is a good deal higher); it has to be matured for at least two years.

Bournville

Bournville is a toponym not only for chocoholics with a taste for something less bland than Dairy Milk, but also for social reformers, economic historians and town planners. John Cadbury, who had started trading as a tea and coffee dealer in Bull Street, Birmingham, in 1824, went into cocoa products seven years later. Trade expanded and the decision was taken eventually to move out of Birmingham which was congested, expensive and unhealthy. A site was acquired four miles south-west of the city centre. It had no particular name, but was called Bournville after the Bourn – a trout stream that passed through – with 'ville' added because it was thought it would give an appropriately French flavour to a confectionery factory. Like some other celebrated chocolate dynasties, the Cadburys were a Quaker family and accepted readily the social responsibility of employers, confident that treating the work force decently would in fact make it more efficient. In addition to building the factory which has been developed over the years, the Cadburys built the Bournville Estate, a model village designed by W. Alexander Harvey and devoted, in the words of the trust deed, to the 'amelioration of the working class and labouring population . . . by the provision of improved dwellings, with gardens and open spaces'.

As well as Bournville Chocolate, there is of course Bournvita, a bedtime drink.

Bren

Bren guns were toponymous twins. Originally manufactured at the Czechoslovakian city of Brno, the British government subsequently had them made under licence at the small-arms factory at Enfield. The first parts of the two names were combined and the British soldier acquired not only a weapon which served him well, but a name around which he could get his tongue and tonsils. Another Second World War weapon named after a place is the Bofors gun, from the industrial town of Bofors in Sweden, north of Lake Vaner. The second syllable of Bofors means 'waterfall'; thus the link with, say, Dungeon Gill Force in the Lake District is plain.

Broderie anglaise

Just what genteelism led to translating 'English embroidery' into French remains a mystery, and in fact broderie anglaise has also at different times been known as Ayrshire, Madeira and Swiss work – presumably indicating different places where it was made – and also as Eyelet work, which gives some indication of the style. The feature of broderie anglaise is that the material, almost always white linen, has small holes – either geometrical in outline or else representing conventionalized flowers – punched or cut in it with a stiletto; these are then sewn round to prevent fraying and produce a decorative effect. Fine broderie anglaise was produced towards the end of the eighteenth century and the style is quite common, even if the work is now done by machine.

Bunkum

Bunkum, as applied to words that don't make much sense, has a history which Henry Ford, thinking time was money, would no doubt write off as bunk. In 1820 at the Sixteenth Congress of the USA, the representative of a district in North Carolina insisted on making a long speech on the Missouri question. Everyone got bored, but he insisted on continuing because he thought it was his duty to make a speech on behalf of his constituency, Buncombe. At first English writers thought that bunkum was an American product, but by 1850 they had learnt better. 'Debunk' was invented around 1923 and has done yeoman service ever since.

Calico

Calico was a name for quite a wide variety of different kinds of cotton cloth imported from the East as early as Elizabethan times, and there was even a tendency to spell it out in full as Calicut. This is of course the name of the town on the Malibar coast of India which became a trading centre from the time of the earliest European contacts with the region, beginning with the arrival of Pero de Covilham and Vasco da Gama at the very end of the fifteenth century.

Calvados

Apple brandy is made in many parts of the world where orchards yield good crops, but real Calvados comes from the department of Calvados in Normandy. This is the part of France which leapt into the headlines when the Allies landed there in June 1944 to embark on the liberation of Western Europe. Apples have been grown in Calvados since at least the eleventh century and in 1553 the first still was set up in Mesnil-au-Val, but it was not until the French Revolution that the name 'Calvados' began to be used for the product. Nowadays Calvados distillation is carefully regulated. It is made by twice distilling a fermented mash of apples and is matured for 6–10 years in oak casks, it is not the same thing as Eau de Vie de Cidre, made by distilling cider. As well as being savoured at the end of meals, Calvados is traditionally drunk in the *trou normand*, the 'Norman gap' – the pause in the middle of the gargantuan meals relished in the region, when the feasters are in need of a pick-me-up.

Camberwell Beauty

Camberwell is part of south London, with Southwark and Bermondsey to the north and Deptford and Lewisham to the east. It might seem an unlikely area for discoveries in natural history, but it was there amidst the willows growing in what was then a village that the first specimens of the Camberwell Beauty were taken in the middle of August 1748 in Cool Arbour Lane; in Latin it goes under the name *Nymphalis antiopa*. The Camberwell Beauty is not a British breeding species; in late summer it comes over from the Continent, usually from Scandinavia rather than France or Belgium, which leads to its being found more often on the east than on the south coast. This is a large butterfly whose wings are brown for the most part, with borders of blue spots and then rich yellow. Entomologists are interested by the way its wing patterns vary according to different temperatures.

Camembert

Camembert, the cheese you have to psychoanalyse if you are to catch it at just the right moment between something with the consistency of plasticine and something that is like a smelly runny egg, is supposed to come from Normandy; the original cheese came from Camembert, a village in the department of Calvados in Normandy. Marie Harel, whose husband had a farm there, is credited with evolving a method of making the cheese around the end of the eighteenth century, and according to legend she learned the secret from a priest whom she concealed during the Revolution. It was not put into round boxes until the 1890s, the beastly practice of wrapping even smaller portions in metal foil being of even later origin.

Canaries

The Canary islands, Spanish possessions since the fifteenth century, lie in the sun out in the Atlantic some sixty miles off the African coast. They have given us a dance, a wine and a bird out of which a football team has been hatched.

In the sixteenth and seventeenth centuries the canarie was a popular dance in triple time; it is said that each partner in turn would dance in front of the other, imitating the gestures of savages. Canary wine, which was known in England in Shakespeare's day, is light and sweet but production took a severe knock when disease struck the vines in the mid-nineteenth century. According to Cyril H. Rogers, the birds we call canaries were not in fact natives to the islands but were introduced there accidentally. A sailing ship with a number of cages on its deck containing birds captured in Africa was sailing towards Leghorn when it foundered off the Canary Islands. Some kindly soul released the birds rather than let them drown, and several managed to flutter to the shore where they quickly established themselves. From there they were exported to Europe where they became popular cage-birds, and it is said that they were brought over to England by Huguenot refugees in the late seventeenth century. Many Huguenots settled in Norwich and the city became famous for its canary breeding. Norwich City Football Club became popularly known as the Canaries and adopted a yellow strip. At first they played at a ground called The Nest, but in 1935 moved to the grander but less homely setting of Carrow Road. After decades in the Third Division, the Canaries have had a run in the First.

Cannibal

Coming into Guiana, the Orinoco region and the Windward Islands, Columbus found himself confronted by a warlike people whom he called 'Caribs', which is taken to mean valiant or daring. As well as giving rise to the geographical term Caribbean, the word has also evolved into cannibal, in recognition of the fact that these tribes ate their fellow men. Suggestions that there might be some connection between cannibal and canine are rejected as unscientific. *The Tempest*, on the other hand, offers an authentic variant form in Caliban.

Canter

The canter is a term in riding for the pace between the trot and the gallop. It is a pretty severe docking of the original form, which was Canterbury gallop, suggesting the pilgrims got along at a reasonable speed without going hell for leather. Another tale from the same source relates to Canterbury bells which jingled merrily on the harness of the horses, and the flowers were called after them. The name Canterbury is also applied to music racks, generally fitted with drawers for storage, and to plate and cutlery stands used for supper parties in the eighteenth century. It is plain from a comment by Thomas Sheraton that the latter takes its name from an Archbishop of Canterbury, and probably the former does also. So these two, if entitled to appear here at all, can only squeeze in because the ingenious prelate took his title from his see.

Carbine

A carbine is a firearm like a musket or rifle, but rather shorter. So it was particularly suitable for cavalrymen who might use it either while on horseback or, with somewhat greater likelihood of hitting the target, after dismounting. The word has been in use since the end of the sixteenth century, the earlier form 'carabin' dropping a letter when the reference was to the weapon. Carabiniers made a considerable impression in campaigns in the Low Countries in the 1690s, and the English 6th Dragoon Guards adopted the name of Carabiniers in 1692. The Italian gendarmerie are also known as the *Carabinieri*. There is no doubt that the English borrowed the name 'carabinier' from the French, just as they copied the French military development of the use of the arm. The French in turn had taken the word from Spanish, 'carabin' deriving from 'calabin', with the 'l' turning into an 'r' by a quite common phonetic change. Still we are not out of the etymological wood, because 'calabin' has two possible senses. It might be a Provençal word meaning merely 'weapon'. However, there is a greater likelihood that the origin lies in Calabria, the region of Italy where the innovation was first seen.

Carronades

Everyone who has ever thrilled to the adventures of Horatio Hornblower knows carronades as the stubby cannons which hurled heavy shot at the ships of Napoleon's navy at short range with devastating effect. They were invented in 1759 by General Robert Melville, who gave them the appropriate name of 'smashers'. But when the British Navy adopted the weapon twenty years later it was called the 'carronade', after the place where it was manufactured – the Carron Iron Founding and Shipping Company, whose works are on the banks of the River Carron near Falkirk in Stirlingshire. In the close-quarters battles of Nelson's time carronades were highly effective, but when the American ships adopted long-range tactics in the War of 1812 the limitations of the weapon became only too obvious.

Charleston

The Charleston, a fox-trot enlivened by vigorous side-kicks, is a dance that summed up the frenetic excitement of the 1920s. James P. Johnson started the vogue with 'Charleston, South Carolina' in 1923. The song and the dance take their name from the cosmopolitan city of Charleston, which its founders named Charles Town in 1670 in honour of Charles II.

Chartreuse

Chartreuse claims a place of honour in any list of toponyms. St Bruno, who was born in Cologne and who – thanks to advertising – may one day become regarded in the popular imagination as the patron saint of pipe-smokers, fell out with the ecclesiastical authorities over what he regarded as their laxity. In 1084 he founded the austere contemplative order of the Carthusians. The name came from the small village in the mountains some twelve miles north of the French city of Grenoble, in the present-day department of Isère, and the monastic site there became known as La Grande Chartreuse. The order flourished and St Hugh, who resided there from 1160 to 1181, founded the first Carthusian house in England at Witham in Somerset, before being translated to Lincoln where he was responsible for much building work in the cathedral and left a reputation for piety. When Henry VIII was imposing his will on the Church in the 1530s the Carthusians were the only religious community to be unwavering in their opposition, and in 1535 the prior of the London Charterhouse and several of his monks were cruelly executed. After the dissolution of the London Charterhouse, the premises passed through various hands until they were acquired in 1611 by Thomas Sutton, a successful businessman by whose will the Charterhouse became an almshouse and a school. In the course of time it became one of the most distinguished of British public schools and is now sited not in London, but at Godalming in Surrey.

After this tale of plain living and high thinking, it would be a come-down to think about the famous liqueur, were it not for the fact that it really is remarkably good. There is no need to go back to St Bruno again but Chartreuse, at least in some early formulation, appears to date back as far as the early seventeenth century. Brother Jérôme Maubec devoted much effort to deciphering ancient recipes and perfecting production methods, and just before he died in 1762 he dictated a set of detailed instructions for the original green liqueur. After surviving the Revolutionary period by various shifts and stratagems, production of Chartreuse began again and Brother Juno Jacquet developed a new version of the liqueur, which was yellow and rather sweeter and less strong than the original version. Not even the expulsion of the Order from France in 1903 caused more than a hiccough – if that's the word – in the success story of Chartreuse, because installations were soon built for making it at Tarragona in Spain, and the Carthusians were eventually able to return to their age-old home. One of the finest liqueurs, Chartreuse has been paid the compliment of many imitations. 'Chartreuse' also gave us the name for a pale apple-green colour; its use is first recorded in 1884.

Cherry

Cherry – like its fashionable French cousin, the colour cerise which is recorded in English in 1858 – has its roots in the Black Sea town of Kerasund (or, as it used to be called, Cerasus). It was from there that Lucius Licinius Lucullus, who was given the additional title of Ponticus because of his great victories in the area but is better remembered by posterity for his devotion to luxury and fine food, introduced the wild cherry into Italy around the year 68 BC. According to Pliny, the cherry quickly spread north, reaching Britain within a century. There is also a rare earth called Cerium, but this has nothing to do with cherries in any form; it comes from the planet Ceres.

Cheshire

Cheshire cheese causes no problems, but etymologists are unable to account for Cheshire cats, which are unknown to breeders and have only one delightful characteristic. Perhaps their grin is the feline equivalent of the Mona Lisa's smile. Alice complained that the Cheshire Cat appeared and vanished so suddenly that it made her quite giddy. So 'this time it vanished quite slowly, beginning with the end of the tail, and ending with the grin, which remained some time after the rest of it had gone'.

China

Chinese lanterns, whether made of paper or grown in the garden; Chinese white, which is a pigment; Chinese tortures, which still curdle youngers' imaginations, and even China tea – all these have had less impact on our vocabulary than china which, losing its capital letter, has been the common English name for porcelain since the seventeenth century.

Chinoiserie

Chinoiserie was the French word used throughout Europe in the eighteenth century to describe objects and designs either from China or in more or less fanciful versions of Chinese style. The new patterns provided a welcome addition to the prevailing rococo tradition. William Chambers' more-or-less Chinese pagoda in Kew Gardens is one of the more impressive monuments to a style usually preserved in smaller items such as Delft figures and vases and black lacquered cabinets with Chinese scenes.

Chinese woodblock

In French, German and Italian, composers are content to call this relatively artless instrument the 'wooden block', but English usage prefers to insist on its Chinese origins. It consists of a block of teak or other hard wood with one or two long cavities; when struck with a drumstick or other beater, it gives out a resonant and penetrating tone. First introduced to the West by ragtime and jazz bands, as the tap box or clog box, it was taken up by such modern composers as Sir William Walton in *Façade*. Presumably renaming it Chinese woodblock was a way of showing it had gone up in the world.

Chivvy

'Chivvy' seems to be the most popular spelling these days, though the dictionaries also show it with just one 'v' and remind us of the variant form 'chevy'. That points to the probable origin of this verb meaning to drive along without respite. The Cheviot Hills in the border country between England and Scotland were the scene of many desperate battles, and the well-known sixteenth-century ballad of Chevy Chase recalls one of them with a catchy rhythm that conveys all the excitement of pursuing and harrying the foe as they scuttle off.

This fight did last from break of day
Till setting of the sun;
For when they rang the evening bell,
The battle scarce was done.

. . .

Of fifteen hundred Englishmen,
Went home but fifty-three;
The rest in Chevy Chase were slain,
Under the greenwood tree.

Coach

Unlikely as it may seem, coach comes from Hungary – originally as carts built at the village of Kocs in the fifteenth century. The word quickly travelled over all Europe with relatively little change in form.

Cognac

Lexicographers take a back seat when legislators are called in on the definition of a term involving such sums of money as cognac. Brandy is made in many parts of the world by distilling wine and then taking various steps to ensure that the fiery spirit produced is potable and consistent in quality. Cognac is a small town in France, a few miles north of Bordeaux, and is the centre of the wine-growing districts from which alone come the grapes from which true Cognac may be made. There are seven districts; in ascending order of reputation they are Bois communs dits à terroir, Bois Ordinaires, Bons Bois, Fins Bois, Borderies, Petite Champagne and Grande Champagne. The term 'Champagne' in this context has nothing to do with bubbly, which comes from a different part of France. Armagnac, which it is perhaps risky to name in the same breath as Cognac, comes from the department of Gers, south-east of Bordeaux; Marc, made by distilling out the grape pomace after the juice for wine has been pressed out, is associated especially with Burgundy, hence Marc de Bourgogne.

Colorado beetle

Colorado beetles are yellow, with natty dark brown stripes down their backs. In the summer you can see pictures of them in nearly every post office and police station in country districts in south-east England, because they are among the most feared agricultural pests and can ruin potato crops. When first discovered in central parts of the United States in 1823, *Leptinotarsa decemlineata*, to use the Latin name, was a beetle of no particular importance which fed on buffalo burr. However, when the early pioneers moved into the region around 1859 and started cultivating potatoes, the Colorado beetle found a food very much to its palate. Not only did it thrive but it began to spread eastwards, reaching the Atlantic seaboard in 1874. There were a number of instances of its being accidentally introduced to Europe, but no great harm was done until control measures broke down completely at Bordeaux in 1920. After that the Colorado beetle spread first across France, then into Spain, Italy and Germany and further east. Only constant vigilance has prevented this beetle and its voracious larvae from devastating British agriculture.

Copper

Copper comes from the Latin *Cyprium aes* – that is, the metal from the island of Cyprus, which was a major source in ancient times. Alchemists represented copper with the same sign as the planet Venus, presumably because Cyprus was particularly associated with the worship of the goddess Venus.

Cor Anglais

Cor Anglais is a toponym which no one seems able to explain. Look up the standard reference works and you will find the instrument described clearly enough as the alto of the oboe family, sounding a fifth lower than the oboe proper. *Oboe d'amore* came in between. The ancestor of Cor anglais is first met with in Germany in about 1720, when it was known as the *Wald-hautbois* or *Jagd-hautbois*, that is the forest or hunting oboe which some composers, including Bach, turned into Italian as *oboe da caccia*. Around 1760 that name dropped out of use and Gluck and Haydn knew the instrument as the *corno inglesi*, which led then to the French form of cor anglais. The 'cor' part of the name is probably explained by the connection with hunting and earlier versions of the instrument were curved in shape. But where the adjective 'anglais' came from remains a mystery, though few nowadays take seriously the suggestion that 'anglais' is a corruption of 'anglé', which some used to suppose referred to the shape of the tube that leads the wind from the reed to the body of the instrument. According to Philip Bate in *The New Grove*, the frenchified form cor anglais is now succumbing to American pressure and is being displaced by 'English horn', but it is questionable whether change is taking place as rapidly as he supposes. Despite the problems over nomenclature, the Cor anglais is immediately recognizable because of its satisfyingly egg-shaped bell and because of its rich tone which Sibelius, for instance, uses so effectively in 'The Swan of Tuonela'.

Cortina

The first Cortina rolled off Ford's production lines in 1963. As one of the most successful cars ever made, it deserves to stand for what could be a whole chapter on models which take their names from places where clever sales executives know drivers would prefer to be while in reality they wait in traffic jams at Watford or wherever. Morris recalled its origins with the Oxford, and Austin riposted with the Cambridge, but both – like the Anglia – lacked the poetry of Grenada and Sierra. Rolls-Royce like to suggest that owners of their cars take their holidays in the south of France, but when they announced the Camargue in 1975 there were protests in *The Times* about naming this prestige-product of British technology after what was described as an area of French marshland. The complainant continued 'have we no British bogs to commemorate?'

Cos lettuce

Cos lettuces trace their ancestry back to the island of Cos, or Stanko, on the south-west coast of Asia Minor at the entrance to the Bay of Budrum; it has been famous for hundreds of years for its agricultural produce. For two centuries after 1315 Cos was occupied by the Knights of St John; they were thrown out by the Ottoman Turks in 1532, but it seems possible that they brought seeds of the local lettuces with them and that by the time of Charles II these were being grown in England.

Cravat

Cravat, in older fashions a necktie and now a short scarf which modern beaux tuck raffishly into their open-necked shirts, comes from Croatia, in modern Yugoslavia. In the Thirty Years War (1618–48) Croatian soldiers wore scarves around their necks rather than stiff stocks.

Cretonne

The strong fabric with a thick warp of hemp or linen and a lighter weft, which is still very popular for chair covers and hangings, is claimed by some authorities to be a toponym derived from the village of Creton, in the department of Eure in Normandy. Others, however, ascribe the name to a linen manufacturer by the name of Creton who, according to Pamela Clabburn's *Needleworker's Dictionary*, introduced the product in 1825.

Curaçao

Curaçao is an island in the Dutch West Indies lying some forty miles off the north coast of Venezuela. It is best known for the variety of oranges which grow there, and from their dried peel the liqueur curaçao was made. There are several types of curaçao, but the brand leader is Cointreau, produced since 1849 by an Angers firm which imported the orange peel from the West Indies through the French Atlantic port of Nantes. The Bols Royal Distilleries at Amsterdam – which trace their lineage back to Lucas Bols of 1575 – produce a number of interestingly coloured curaçaos. Among them is a blue variety which certainly makes drinks eye-catching – though it is hard to say what, if anything, the hue does for flavour. In the 1930s blue curaçao was the essential ingredient in the 'Marina blue' cocktail created after Princess Marina of Greece married the Duke of Kent in 1934. Blue Lagoon is made by pouring a measure of blue curaçao over cracked ice and topping up with lemonade; it looks just right for a blazing hot day on the beach, but to my palate it is a bit too sweet.

Currants

Currants, among the best known of toponyms, came by way of the French and Middle English 'raisins de Corauntz' from Corinth. The Greek city also gave us Corinthian capitals, which with their rich decoration of stylized acanthus leaves confer on the grandest public buildings an air of confident prosperity. Over the years the word has also developed from a term of disapproval for opulent living tinged with degeneracy and even effeminacy – see St Paul's comments – into an expression of respect for the highest standards of sportsmanship amongst the upper classes. The Royal Corinthian Yacht Club was founded in 1872 for gentlemen who sailed their own boats, not employing professional skippers. Corinthian AFC, which amalgamated with the Casuals in 1939, was founded by N. L. Jackson in 1882 and with its teams recruited from former public school players, became the most famous amateur soccer club in the country.

Dalmatians and Dalmatics

Dalmatians – moderately large dogs with a profusion of black or brown spots on their white coats – appear in Greek and Roman art, but the modern breed seems to have originated in the eighteenth century in Dalmatia, the country on the Adriatic coast of modern Yugoslavia. In the eighteenth century it was the height of fashion to have a Dalmatian running along under your carriage, which provided left-wing orators with their parrot cry about 'running dogs of the capitalist imperialists' when describing their opponents.

A dalmatic is a liturgical vestment, a tunic with square-cut sleeves and tail and with broad stripes running down from shoulder to hem. Originating from Dalmatia, it came into fashion in the second century AD and when first adopted by the early Christian church, was reserved for those of the clergy who were in a special relationship with the papacy. In the course of time it became more widely used, especially by the deacon and sub-deacon at high mass. At coronations the monarch is robed in a dalmatic as a symbol of quasi-priestly status.

Damsons

Damascus has fathered several toponyms. Damsons were first encountered as Damascene plums, which was doubtless too long a name for such small fruit. Damask is the rich fabric – first silk and subsequently other kinds – with patterns and figures woven into it, while damask roses were thought to resemble the cloth. In metalwork, especially the ornamentation of weapons, damascening is surface ornamentation with inlaid patterns, generally of gold or silver.

Dartford warbler

Dartford, currently celebrated for its traffic jams at the toll-booths at the southern end of the Dartford Tunnel, enjoys a certain ornithological fame for the Dartford warbler, a toponym enshrined in the Latin title *Sylvia undata dartfordiensis*. The first recorded sightings were made at Dartford and in nearby Bexley in 1773 by John Latham (1740–1837), who practised medicine at Dartford some years before retiring to devote his energies to bird-watching and the writing of important early works on ornithology. The Dartford warbler is an alert little bird, about five inches long; it is coloured greyish-brown on top and rust brown below, with a speckled chin, and there is a fine reproduction of Lilford's 1887 coloured plate in James Fisher's *The Birds of Britain*. The French name for the bird is 'pitchon', which is an onomatopoeia for its call. Building and development have destroyed the Dartford warbler's favourite habitat, gorsy heaths, so it is rarely seen near the town; but it is still found in rural parts of south-east England, in Brittany and the Atlantic provinces of France and in Spain.

Delft

An old town in Holland about five miles inland from The Hague, Delft has donated its name to earthenware which is given a tin-glaze and generally decorated in blue, yellow or green with vigorous brush-strokes. The origins of the technique and style go back as far as Mesopotamia. Spreading to North Africa, it was brought over into Spain by the Moors. It had already acquired the names majolica (from Majorca) and faience (from Faenza, in northern Italy) before being adopted by Delft. Still its wanderings were not over, for a pair of Dutchmen brought it over to Norwich in about 1570 and pottery in the same style was to be manufactured in London, Bristol, Liverpool, Glasgow and Dublin.

Demerara

Demerara rum is of course casually connected with Demerara sugar, and both borrow their name from the Demerara river which flows north through what is now called Guyana to reach the sea by Georgetown. Rum is made by distilling the molasses left after the characteristic brown Demerara sugar has crystallized out; at least in some degree, the dark colour is produced by adding caramel. Until the recent fashion for white rums, Demerara rum was the most popular variety in Britain, followed by the lighter coloured and less pungent rum made in Jamaica.

Denim

The beauty of denim as a toponym is that it not only contains the name of the place where the product originated but is preceded by the word 'de', i.e. 'from'. In other words, denim means cloth 'from Nîmes', the city in southern France between Avignon and Montpellier. The name has stuck even though the cloth seems to have changed. Once denim meant a sort of serge, but now it is cotton twill.

Dollars

Few toponyms have had wider currency than dollars. The story begins in 1518 with the minting of large coins from silver extracted from the mines in Joachimsthal – the Vale of St Joachim, the father of the Blessed Virgin Mary – in Bohemia. 'Joachimsthaler' was too long a word to survive for long, but 'thalers' and then by a quite normal phonetic shift 'dollars' caught on throughout Europe and then spread to the New World. In the USA the dollar was authorized by Congress in 1792, its value being fixed as equivalent to that of the Spanish milled dollar; the first silver dollars were issued two years later. Maria Theresa dollars struck in silver and bearing the date 1780 continued to be minted until recent times because they were required for commerce in the Levant and, indeed, right across North Africa. When silver was in short supply in Britain at the end of the eighteenth century, Spanish silver dollars were allowed as legal tender; they were countermarked with the Goldsmiths' Company's stamp and circulated with a value of four shillings and ninepence. In 1804 the Bank of England issued two million five-shilling tokens struck on Spanish dollars.

There are disputes about the origins of the dollar sign. The most plausible explanation is that $ is derived from /8/, which was used by the Spaniards to designate the peso or piece of eight which we usually hear about in tales of the buccaneers.

Doric

The Dorians were one of the main groupings of Hellenic peoples. They took their name from a small district called Doris in northern Greece, near Mount Parnassus, but a large area was considered to be 'Dorian', its inhabitants forming a confederacy whose influence was spread all the wider by colonies. By comparison with other peoples of Ancient Greece, the Dorians were regarded as somewhat backward. In the Dorian mode, one of the Greek modes of music, there was felt to be the strength and solemnity befitting people still unsoftened by over-much civilization; while the Doric order in architecture, with its simple capitals, is decidedly more severe than the Ionic and, particularly, the Corinthian orders. The emphasis falls squarely on the country bumpkin side when 'the Doric' becomes a half-apologetic euphemism for the local patois, especially in Scotland.

Duffle

Duffle coats are made of a warm but coarse woollen cloth which originally came from Duffel, a town in Belgium a few miles from Antwerp. The spelling Duffle seems to be winning, presumably because it chimes with English usage generally and 'muffle' in particular. Released by the Navy as surplus stores at the end of the Second World War, duffle coats with wooden toggles and a cut that concealed the sweater worn underneath almost became civilian uniform in the chilly winters of the late forties. Since then duffle coats have gone up in the world of fashion and in price. Duffle bags were originally made of thick cloth too, but now the term is applied to any cylindrical canvas bag carried by the drawstring which closes the opening at one end.

Dum-Dum bullets

'Dum-Dum' were hollow-nosed bullets which expanded on impact and caused a large, ugly wound. They were developed in India for use in the campaigns on the North-West frontier because ordinary rifle bullets would not stop a charging tribesman unless they killed him outright. The name comes from Dum-Dum, a cantonment in Bengal a few miles north-east of Calcutta which in Victorian times became the most important arms and ammunition factory in British India. The British authorities were most upset by suggestions that dum-dum bullets were used against white men in the Boer War, though they were not keen to accept proposals made at the Second Hague Conference that their use should be totally banned.

Dunstable

To say something in Plain Dunstable or Downright Dunstable used to mean saying it without beating about the bush. However, it is doubtful whether this has anything to do with the speech habits of the Bedfordshire town. Probably it comes from the phrase 'plain as the road to Dunstable', which could be a reference to the fact that the route from London to that town was straightforward, but then forked off in various directions. But a pun on 'dunce' may have added something to the topographical allusion.

Encyclopaedia Britannica

Some of the most celebrated reference works are known by titles recalling their place of origin rather than the names of their authors or editors. *The Dictionary of Phrase and Fable*, that invaluable compendium of allusions which we never care to admit we cannot quite place, is forever associated with its compiler, E. Cobham Brewer, while successive editions of the *Thesaurus* still bear the name of Peter Roget. Larger projects present themselves more impersonally. The *Complutensian Polyglott*, for instance, is one of the monuments of Biblical scholarship, giving the Scriptures in their original tongues together with a Latin translation. This vast work, which was begun under the direction of Cardinal Ximenes and completed in 1517, was printed in the Spanish city of Alcala, known in Latin as Complutum. The original and self-explanatory title of Sir James Murray's monumental contribution to lexicography was *A New English Dictionary on Historical Principles*; subsequently it became *The Oxford English Dictionary*, marking the support given by the Oxford University Press to the huge enterprise. Probably two points were being made when the title of *Encyclopaedia Britannica* was allotted to a new reference work which first appeared, in Edinburgh, in 1768. In the first place, there was an implication of comparison with the French *Encyclopédie*, the 'Reasoned Dictionary of Sciences, Arts and Trades' whose publication in the third quarter of the eighteenth century was one of the major manifestations of the critical spirit of the Enlightenment in France. And in the second place, 'Britannica' insisted that this was not an English but a Scottish production. However, over the years the encyclopaedia gradually lost its connection with Scotland. The fine eleventh edition which came out in twenty-nine volumes in 1910 and 1911 appeared under the aegis of the Cambridge University Press, which nevertheless had nothing to do with its preparation and printing. In reality, by then the *Encyclopaedia Britannica* had been taken over by American publishers from whom it eventually passed into the care of the University of Chicago.

Epsom salts

Before becoming famous for the Derby – the world's best known horse race which was first run in 1780 and takes its name from the Earl of Derby – and for the Oaks – instituted a year before and called after his nearby country house – Epsom was known as a spa. Mineral springs were discovered there in 1618 and a period of great prosperity followed, with Charles II frequently taking the waters. Prince George of Denmark, Queen Anne's consort, was a regular visitor too, but as the eighteenth century passed other spas became more popular. Epsom salts is a familiar name given to magnesium sulphate, the purgative found in sea-water as well as in most mineral waters including, in addition to those discovered at Epsom, those drunk at Seidlitz and Pullna.

Fiat

Fiat, the name of the car and truck marque, is an acronym that is 50 per cent toponymous; the letters stand for *Fabbrica italiana automobili Torino* (the Italian automobile works at Turin). The firm was founded in the northern Italian city in 1899 by Giovanni Agnelli and, although still mainly identified with vehicle construction, has substantial interests in many other branches of engineering too. For the British market the word 'Fiat' was well-chosen; it is short and simple and the legal term 'fiat' – meaning an instruction or petition emanating from a high legal officer – provides a pattern for pronunciation without any danger of confusion over meaning arising, except perhaps in chambers occupied by barristers who are motoring enthusiasts! For Italians, however, 'Fiat' is a little too abrupt, with an uncomfortable final consonant; generally they contrive to add a concluding vowel.

Florence Nightingale

Many surnames are of course toponyms and peers proclaim their territorial associations. Military and naval commanders even pin the place of their most notable triumphs to their titles, so that we have Alexander of Tunis and Montgomery of Alamein. Among the most famous of first names chosen to recall where the child was born is that of Florence Nightingale. Her father, a well-to-do and cultured gentleman from Derbyshire, had a taste for travelling and when his first daughter was born in Naples, he called her Frances Parthenope, choosing the latter because that was the city's Greek name. He went one better when Florence was born in Florence on 12 May 1820. The name had been used in England before then, for men as well as women, but Florence Nightingale's achievements gave it great popularity for girls until it succumbed to the effects of abbreviation – to Flo, Florrie and, worst of all, Flossie.

Florins

The first florins were struck in Florence in the middle of the thirteenth century. On one side these gold coins had the head of St John the Baptist and on the other the lily (the *fiorino* or little flower) which was that northern Italian city's emblem. Together with the ducat from Venice, the coin set the fashion for European mints and in 1344 Edward III issued golden florins which had a value of six shillings. In Victoria's reign there was an early movement towards the decimalization of British currency, and the florin was the name given to the coin with a value of one-tenth of a pound (i.e. two shillings); this was intended to replace the half-crown, worth two shillings and sixpence or an eighth of a pound. The first of these silver florins was introduced in 1849, but the Queen's name was not followed by the customary 'Dei Gratia, F.D' (by the Grace of God, Defender of the Faith). There was a popular outcry against these so-called godless coins, an outbreak of cholera was even taken as a sign of divine displeasure and the issue was recalled. Despite this setback, the florin did displace the half-crown between 1852 and 1874 and thereafter the two coins existed together until the half-crown disappeared with the advent of decimalization. The double-florin or four-shilling piece had only a short life, from 1887 to 1890.

Frankfurters

Frankfurters, those pink sausages which come in pairs and are cooked in boiling water, have impeccable German connections but have had a less exciting life than hamburgers. The latter no doubt had links with the great German seaport, but that first syllable tickled the fancy of the ad-men. Hamburgers do not contain ham, but you soon face trouble if you hint that beefburgers are not 100 per cent beef, while cheeseburgers drip with cheddar. Perhaps the hardest to swallow, linguistically, is the TV-burger!

French

Over the centuries of rivalry and friendship, France has given English a hamperful of toponyms. 'Say Noilly Prat and your French will be perfect' was an advertising slogan which made fun of our embarrassment about getting our tongues around French words. French kisses bespoke habits we were not quite sure about either, and in the Renaissance the French declared they were Italian. French cricket is played without stumps, so if the ball touches your legs you must be LBW; and you don't need needles for French knitting. French polish, though, is a tribute to French furniture, while French windows – sometimes so naturalized as to be written without a capital letter – have added to the amenity of many a house with a pleasant garden. French marigolds are bushy and exist in many fancy varieties; they are altogether more delicate than the common type. French dressing equally deserves our gratitude for what it has done to help salads slide down more easily, and the Americans insist on calling chips 'French fries'. Instead of merely putting dry bread under the grill, French toast is made by dipping the slices into lightly seasoned beaten egg and then frying them. French chalk was used by tailors for marking cloth, and French drains are trenches filled with rubble which sometimes take the water off the surface of car parks. Looking for someone to blame, our ancestors called the pox the French disease and, apparently with reference to the French practice of leaving receptions without formally thanking one's host and saying farewell, they liked to use the scornful euphemism 'to take French leave' when they meant that their enemy had run off. The French had their revenge with *filer à l'anglaise*, which means just the same. And, in case you are wondering, there is a French town called Condom; it is a small place in the department of Gers, in south-west France, about thirty miles north west of Auch, and is a centre for the sale of Armagnac.

French horn

Among classical musicians there is a growing tendency to call the instrument simply the 'horn'. This could give rise to confusion with the terminology of the jazz band, but nobody seems worried about that. In the eighteenth century horns by French makers did set something of a standard, and the name became fixed in English despite developments in the design of the instrument in Germany.

Fustian

Fustian has come such a long way in both meaning and geography that it really is a shame that it seems always to attract the dismissive adjective 'mere' when used nowadays in the figurative sense. After taking Cairo in AD 641 following a year-long siege, Amr founded the town of El Fostat nearby; he gave it that name, meaning 'the tent', because that was where he had pitched camp. From El Fostat came coarse, thick cloth made of cotton and flax. Even by Shakespeare's time fustian had developed into a byword for turgid bombast and ranting, especially of the theatrical variety.

Galilee

Galilee, with obvious connotations of the Holy Land, is the attractive name sometimes used for a porch or else a chapel at the entrance to a church. One of the most attractive galilees in England is to be seen at Durham Cathedral, the comparative lightness of its Transitional columns contrasting well with the sturdy Norman pillars of the nave as you pass through.

The French dictionary-maker Emile Littré thinks it at least worth recording the speculation that the word 'gallery' may be linked with Galilee too, most probably through the transferred architectural sense. *O.E.D.* appears quite unimpressed by the idea.

Gauze

Though the lexicographers are tentative, it seems likely that this thin and virtually transparent cloth takes its name from Gaza, a city famous in Scripture for some of Samson's exploits. Did the harlot he visited there revealingly swathe her charms in gauze?

Gavotte

Many of the folk dances which were given a patina of refinement and admitted to classical music have names linking them with places. The elegant Sicilienne in 6/8 time began, it is said, as a shepherd's dance in Sicily; while the Allemande, though sometimes thought to have started its life in the Low Countries, proclaims a connection with Germany. The Gavotte is so called because it originated among the Frenchmen in the area around Gap, in the Hautes-Alpes department, who were known as Gavots.

German measles

Germanium is a metallic element which was discovered in 1886 by C. Winkler who named it after his fatherland, while German silver is an alloy of copper, nickel and zinc originally found in an ore mined at Hildburghausen. But German measles, though generally written with a capital letter, is a disease related (or germane) to measles and has nothing specific to do with Germany. *Rubella*, to give it its scientific name, is a disease which is mild in itself but can have serious effects on the foetus if the sufferer is pregnant.

Gorgonzola

It will come as no surprise to anyone to learn that Gorgonzola, the cheese that makes an immediate impact, comes from a place of that name. That it is a small town in Lombardy, just a few miles east of Milan, may not be quite such common knowledge.

Greek fire

In two great sieges, 'the deliverance of Constantinople may be chiefly ascribed to the novelty, the terrors, and the real efficacy of the *Greek fire*.' So wrote Edward Gibbon in *The Decline and Fall of the Roman Empire*. Yet he also had to admit that the new weapon, unlike the invention of gunpowder, did not constitute a complete revolution in warfare, which was one of the many reasons why the Empire in the East eventually fell. Inflammable compounds of various kinds had not unnaturally been used in sieges since the earliest times, and their exact composition – which generally included sulphur, pitch, charcoal and sometimes naphtha – was always a closely guarded secret. The improvement which so greatly helped the Byzantine armies and navies was the discovery of a means of making it catch fire apparently spontaneously. This was probably done by including in the mixture a proportion of quicklime; this generated a lot of heat on coming into contact with any water and caused the other ingredients to flare. According to Gibbon, the inventor of Greek fire was one Callinicus, an early chemist from Heliopolis in Syria who deserted the caliph and placed his invention at the service of the Greeks. Whether or not all this is true, it does serve as a reminder that chemistry was an object of special study by the Egyptians, from whom the Arabs learned a great deal long before much was known about the subject in Western Europe.

Gruyère

Gruyère cheese came originally from the Gruyère district in the south-eastern part of the Swiss canton of Fribourg, which itself lies between Lake Geneva to the south and the Lake of Neuchâtel to the north. The district is so called because the castle there was the seat of the counts of Gruyère; they are first mentioned as early as 1073, and their name is supposed to be derived from a word meaning an officer of woods and forests. The temptation of punning heraldry proved too great, however, and the family's coat of arms bore the device of a crane (in French, a 'grue'). The district is not only famous for that splendid cheese with holes in it; it was also the home of the so-called 'Ranz des Vaches', the herdsmen's melody for calling home their cows at milking time. William Wordsworth wrote a sonnet on hearing this from the top of the St Gotthard Pass during his continental tour of 1820; he commented on the feeling the melody was said to arouse in homesick natives of those parts.

Guinea-pig

Guinea-pigs are in no way related to hogs, but there is no doubt about their connections with Guiana. It is generally accepted that the domestic guinea-pig was brought over to Europe from the northern parts of South America in the seventeenth century. Probably descended from the Brazilian cavy, the species had been domesticated by the Incas. A South American relation of the guinea-pig is the Patagonian hare (or mara), which can show a good turn of speed on its long legs. These animals have been used so often for laboratory experiments that nowadays human beings who take part – willingly or otherwise – in the testing of new schemes or products have also come to be termed 'guinea-pigs'.

Guineas

Like guinea fowl, the gold coin guineas have their links with the region of West Africa that is still sometimes loosely called the Gold Coast; Ghana is an older version of the name which has had obvious attractions in the post-colonial era. The first guineas were coined in 1663, during the reign of Charles II, from gold brought into Britain by the Africa Company whose badge of the elephant (and sometimes the elephant and castle) was stamped on the coins in early years. Spade guineas, so called because the royal arms on the reverse looked like something copied from playing cards, were issued during the reign of George II from 1787 to 1799. The guinea was replaced as the standard gold coin by the sovereign in 1817, but has lingered on until recently as a polite unit of account with a value of one pound, one shilling for professional fees and luxury purchases. In racing – think of the Two Thousand Guineas, dating from 1809 – it will no doubt run for ever.

Guernsey

From Guernsey – from the Old Norman 'Grenezay' or Green Island – have come the cows that were prized for their rich yellow, creamy milk before we all became cholesterol-conscious. The product of crossing two types of French cattle, one of which was introduced into the bailiwick as far back as AD 960, the Guernseys have been pure-bred for nearly two centuries. Fawn, golden and sometimes deep red with flesh-coloured or white muzzles, these cows browse contentedly oblivious of the tourists when tethered to a piquet on a common or by a roadside. In *Cranford* Mrs Gaskell dressed Betsy Barker's Alderney in dark grey flannel, while A. A. Milne turned to one to make 'butter for the Royal slice of bread'. But Alderneys are not recognized as a distinct breed of Guernsey cattle.

The Guernsey lily (*Nerine sarniensis*) is a delicate pink or crimson flower which is the emblem of the bailiwick, though it originated well up the slopes of Table Mountain in South Africa. How it came to Guernsey is uncertain, but most likely the plant – which had been seen in Europe as early as 1634 – was brought to St Peter Port by Sir John Lambert, a Parliamentary general and keen gardener who was imprisoned in Castle Cornet after the Restoration.

Guernsey is renowned for its knitting industry, and Guernseys in traditional style of navy blue worsted wool are worn nowadays by many who would never never go to sea.

Habanera

The habañera, the dance now inseparably linked with Bizet's chromaticism and Carmen's slinky sinuosity, is a measured, sensuous dance in duple time from Havana. No doubt its origins are really African, but it was from Cuba that it came across to Spain in the mid-nineteenth century.

Hafnium

The capital of Denmark has enriched our vocabulary with a curious set of toponyms. Copenhagen is a drink made by whipping up eggs in spirits, and also an artless game, while Copenhagen blue is a rather dull colour. The dreadnought Admiral Fisher, who thought Nelson could do no wrong, used to relish the prospect of 'copenhagening' the German Fleet, by which he meant that it should be destroyed in what we have come to call a pre-emptive strike. Neither the word nor the idea caught on at the Admiralty, though the Japanese twice showed the effectiveness of the tactical stroke at Port Arthur and at Pearl Harbour. In 1923 a newly discovered metallic element was named Hafnium after Copenhagen (or in Danish, Kjöbnhafn), which itself means 'traders' harbour'. Copenhagen was also the name of the Iron Duke's favourite horse; it carried him throughout the long day at Waterloo.

Hanoy Tower

Many toponyms are a reflection of history, but just occasionally they seem to foreshadow the future. As far back as 1933 Hubert Phillips could write that Hanoy Tower is a game of patience that 'is quite unlike all other patiences in one respect – it must come out; it is an exercise of ingenuity rather than a patience.' Basically it is quite simple. Take nine playing cards numbered two to ten; you can perfectly well just scribble the numbers on scraps of paper, but make certain that you write six and nine in words as well as figures or you are sure to succumb to the temptation to cheat as your frustration grows. Shuffle the cards, then distribute them in three columns of three cards each. All you have to do is transfer the cards one at a time, so that you end up with just one column containing all the cards in descending order of value. To accomplish this you are allowed to move only one card at a time, however; this must come from the foot of a column and be placed at the foot of either of the two others, below a card which is of a higher value. If it so happens that you have removed all the cards from one of the columns before the patience comes out – and believe me, it always does! – then you are at liberty to take a card from the bottom of either of the remaining columns, so that you can enjoy the flexibility of working in three columns once again. Whether we can look to the emergence of a variant called Ho Chi Min City Tower is uncertain.

Helium
Helios, son of Hyperion and Thea, was the god of the Sun and Heliopolis – in ancient Egypt at the apex of the Nile delta – was devoted to sun worship, though not in quite the same way as the Costa del Sol. As well as numerous technical terms, Helios gave Sir J. Norman Lockyer the name helium for the inert gas he identified in 1868. The heliotrope was so called because the flowers turned towards the sun to a marked degree; at one remove, heliotrope then became an adjective to describe their colour.

Hessian
Hessian boots, which came high up the calf and had tassels in front, were first worn by troops in the employ of the German principality of Hesse. They were fashionable in Britain in the first half of the

nineteenth century, as is clear from the novels of Dickens. Quite why the coarse but strong cloth made of hemp and jute which formerly was used for covering bales and for the undersides of upholstery – and which a few years back was much favoured by trendy interior decorators – is called hessian, is uncertain, but there can be little doubt about the connection.

Hippopotamus

Can a case be made out for the hippopotamus as a submerged toponym? Taken literally from the Greek, the word comes across as river horse. But does that mean the kind of horse you might find in *a* river or in *the* river *par excellence*, the Nile? It is true that it is not only in the glorious mud of the Nile that hippopotamuses can be found cooling their blood, but the Germans seem to concede the toponym when they translate hippopotamus as *Nilpferd.*

Hock

The word is simultaneously a crude abbreviation and a cynical widening of the original term. Hochheim is a town on the river Main in the Rheingau district of Hesse-Nassau, and Hochheimer was the wine that came from there. Now hock means virtually any white wine from the Rhineland that is not good enough to be identified more precisely.

India

India provides an argosy of toponyms. Indigo is a dark blue vegetable dye from the sub-continent, and dictionaries also speak of Indian red and Indian yellow. Even if originally made with pigments from China and Japan, Indian ink is waterproof and very black; India paper is thin yet opaque, which means the ink does not show through, and India rubbers serve to rub out mistakes made in pencil. Indian tonic water emerged under the Raj as a means of making quinine palatable. Indian clubs were much in vogue for manly physical exercise between the 1880s and the First World War, and how to do the 'Indian rope trick' is still a puzzle. The grand geographical misunderstanding which supposed the New World to be the other side of the Far East gave us first the West Indies and then the Red Indians, and it is from them that we derived Indian corn, Indian file and Indian summer.

Inverness cape

Inverness capes, which seem to have been named after the place rather than because they were first made there, were just the thing for a Dickens of a winter. A capacious loose overcoat that reached down to the knees, it had not only sleeves but, for extra warmth, a half cape in front which falls from collar to waist. Large pockets were sometimes provided too, which made the Inverness cape even more suitable for travellers embarking on lengthy train journeys.

Jeans

When jeans swept into fashion all over Europe it looked like another wave of American influence. In fact, it turns out that Levi's and the rest were only shipping back a word, if not a product, which had been sent out from the Old World. Jean is derived from the Italian city of Genoa and this name for a type of cloth has been around in English since Elizabethan times.

Jerry-building

Jerry-building, as a word, has been in use since the 1880s, although the question of the etymology has been the subject of some speculation. It would be pleasant to think that it is a reference to the Walls of Jericho which, as is recorded in the *Book of Joshua*, came tumbling down probably in the thirteenth century BC. Some etymologists, however, believe the term relates to the building firm of Jerry Bros. which ran up speculative housing developments with shoddy materials and poor workmanship at Liverpool early in the last century.

Jersey

Jersey cattle are similar to their Guernsey counterparts except for being rather smaller and having dark muzzles, but in the sense of being a comfortable woollen garment with long sleeves covering the upper half of the body, jersey has become a far more generalized term than guernsey. The so-called Jersey Lily is the *Amaryllis belladonna*, a native of South Africa. Lily Langtry, the comely actress who moved in the Prince of Wales's fast set towards the end of the Victorian era, was commonly called 'the Jersey Lily' because she was the fair daughter of the Dean of Jersey.

Jet

Jet is a word that now survives as much in the cliché compound adjective 'jet black' as in the hard black lignite from Whitby in Yorkshire that was in great demand in Victorian times for mourning ornaments. The word comes via French from Gagas, a Greek town in Lycia (Asia Minor) where it was mined in ancient times.

Jodhpurs

O.E.D. quotes W. G. Steevens in praise of jodhpur riding breeches in 1899: 'Breeches and gaiters all in one piece, as full as you like above the knee, fitting tight below it, without a single button or strap.' Such a practical garment for riding was bound to catch on, especially with the pony club girls riding astride who have replaced those intrepid ladies who rode side-saddle in voluminous habits. The name comes from Jodhpur, a town in Rajasthan in north-west India.

Jovial

Speculations that human character and behaviour are influenced by heavenly bodies and their conjunctures have given rise to a number of toponyms. A cheerful, good-natured fellow is called jovial because he is supposed to have been born under the influence of Jupiter, the planet of Jove, while somebody who is of volatile temperament is described as mercurial, by reference to Mercury. Those born when Saturn was in the ascendant are said to be saturnine, or gloomy in disposition. Lunatics were so termed because it was imagined that their condition was governed by the phases of the moon, and more generally, disasters – taken literally – are events which occur when the stars are against you.

Kaolin

Kaolin is often more familiarly known as china clay, except in the context of diarrhoea remedies where presumably a little bit of verbal mystery mingled with the morphine is supposed to strengthen the medicine. The name is said to be a corruption of the Chinese Kau-ling, meaning 'High Ridge', which refers more specifically to a hill east of King-te-chen. It was there that samples of the clay, a form of aluminium silicate, were collected in the early years of the eighteenth century by a Jesuit missionary who shipped them home to France for scientific investigation.

Labyrinthine

Some authorities argue that labyrinthine is an eponym rather than a toponym, reasoning that the first labyrinth was constructed for Labyris, an Egyptian monarch of the twelfth dynasty. However, it could be that the word comes from a Greek term for a mining gallery, and another ingenious suggestion is that it refers to a word meaning 'double-edged axe', the symbol of Zeus. However this may be, it is likely that in the western tradition all the many words we have which are connected with labyrinth refer back not to the Egyptians, but to the underground architectural puzzle constructed after the Egyptian example by Daedalus for the imprisonment of the Minotaur in Crete. When Theseus elected to go to Crete as one of the seven youths and seven maidens King Minos demanded from Athens as an annual tribute to be sacrificed to the Minotaur, Princess Ariadne fell in love with him. So she provided him not only with a sword to slay the monster but also with a skein of thread so that he could find his way out of the labyrinth again. Now, as Ivor Brown points out in *Words in Our Time*, that thoughtful idea also gave us the word 'clue'. Nowadays this generally means a more or less helpful suggestion for the solution of a problem which we cannot resolve; originally, however, it meant quite simply a 'ball of thread' and only gradually assumed the figurative meaning which has now virtually hidden the literal one. The original meaning is kept alive as a technical term among seamen, however, under the spelling 'clew' in connection with various arrangements for lacing sails or hammocks. Erskine Childers' splendid tale *The Riddle of the Sands* adds a further twist to this amazing story. When Davies scribbles down some misspelt notes for the guidance of Carruthers, who is rather aware of his own educational superiority, he writes 'clews' for 'clues'. It would be good to know who was having the last laugh in that little encounter.

Laconic

Sparta was the capital of the powerful Greek state of Laconia and its people accepted an austere life-style in order to remain hardy and strong. When one speaks of 'spartan conditions' one is paying tribute, albeit unwillingly perhaps, to the inhabitants of that place even if only to complain about their habits. Spartans were no keener on extravagance in words than in any other sphere, and a laconic style does not use two words where one will suffice. A single example – obviously – must make the point. When Philip of Macedon tried to scare the Spartans he wrote: 'If I enter Laconia, you will be exterminated.' They replied: 'If.' It is a grand story but, alas for those who think style matters, Philip won the war.

Lambeth Walk

The Lambeth Walk became a popular party dance after it was introduced into ballrooms in the late 1930s. The pairs of dancers repeated a basic sixteen-bar sequence of steps as they went around the floor, generally singing the song 'Doing the Lambeth Walk' which echoed on for years until it began to sound like something traditional.

Landau

Landau is a town in Bavaria, some thirty miles south-west of Mannheim. It seems reasonable enough to suppose that it has given its name to those four-wheeled carriages with tops which can be raised or lowered according to the weather – nowadays seen in state processions for occasions or personages one notch down from the most magnificent. However, it has also been suggested that the landau is so called because it was introduced in the mid-seventeenth century, by an English carriage builder with the name of Landow.

Leghorn bonnets and Leghorn fowls

For centuries Leghorn has been the fairly incongruous name by which Englishmen have known the Italian city of Livorno. Churchill had strong views on the matter and in January 1941 he found time to send a memo to the Foreign Office couched in the following terms: 'I should like Livorno to be called in the English – Leghorn; and Istanbul in English – Constantinople. Of course, when speaking or writing Turkish we can use the Turkish name; and if at any time you are conversing agreeably with Mussolini in Italian, Livorno would be correct.' It is clear, however, that the tendency nowadays is away from English forms which are felt to show a lack of proper regard for local sentiment. It has given us two toponyms. Leghorn bonnets were made of plaited straw from the area around Livorno; it was cut while the wheat was still green and then bleached. Leghorn fowls are found in several colour varieties, of which the Brown and the White are the most important. Originating in Tuscany, Leghorns are now found in many countries around the world; they are both hardy and excellent egg-layers. Bantams are small fowl that hark back to a town called Bantam in Java.

Lesbianism

When Byron wrote of 'the isles of Greece, where burning Sappho loved and sung', he was thinking of Lesbos mainly as a centre of high Greek culture. The island lies in the Aegean Sea, off the coast of Mysia, north of the entrance to the Gulf of Smyrna. As a remarkably fertile island conveniently placed on the trade routes through to the Hellespont, it enjoyed great prosperity which reached a peak in the sixth century BC. Sappho, who was born on Lesbos around this time, is generally reckoned the most accomplished of Greek poetesses and is credited with a number of innovations in lyrical technique. The term Sapphism was coined only relatively recently to designate the sexual tendencies she is supposed to have had. The toponym Lesbianism refers to her too, but less directly. It is probably something of a euphemism and seems to have pushed the more precise (if not necessarily entirely accurate) word out of common usage.

Lawn

Lawn, the very fine linen or cotton fabric used for ladies' blouses and archdeacons' surplices, takes its name from the French cathedral town of Laon, in the department of Aisne, ninety miles north-east of Paris. The derivation appears less unlikely when it is recalled that Laon is pronounced in French as if there were no 'o' in the word.

Limerick

Whether limericks have anything to do with the Irish town is doubtful, though one theory has it that the poetic form which has attracted so many wags was brought there around 1700 by soldiers returning from the wars in France. Another suggestion is that it is linked with, 'Will you come up to Limerick?', the refrain of a ditty much favoured by revellers at social gatherings where such verses were greatly appreciated.

Limousine

Limousines have had to travel a long way to become superior motor cars with their cosseted passengers in a compartment separated from the driver by a partition. The term comes from Limousin, the region in central France around Limoges, which is a town famous for its porcelain. In the nineteenth century a 'limousine' was a shepherd's thick, warm cape, sometimes with a hood fitted to keep out the wind. The French word was transferred to luxury automobiles around 1900 and crossed the Channel in 1902.

Lusitania

Many ships are named after places of course, but this is not always very obvious. For instance, virtually every aspect of the sinking of the *Lusitania* by a German U-boat off the Old Head of Kinsale on 7 May 1915 has been examined in detail. It was an act that caused the deepest resentment in America, and suggestions have even been made that the entire affair was engineered by the British Admiralty in order to bring the USA into the war. Amidst the crimes and the controversies it is generally overlooked that the 31,000-ton Cunarder was named after Lusitania, which is the ancient name of roughly the same region as modern Portugal – just as her sister-ship, the *Mauretania*, took her name from a region somewhat further south, the Roman province in North Africa which also has an Atlantic seaboard.

Lyddite

Lyddite is an explosive composed mainly of picric acid. It was first tested on the shingle beds near Lydd, a little town in Kent in the south-east corner of Romney Marsh. For a while it was much used by British forces, but was replaced by TNT (Trinitrotoluene) early in the twentieth century.

Macédoine

O.E.D. has an entry, dated 1846, for macédoine, which it defines as fruit or vegetables embedded in jelly. *Larousse Gastronomique* does not regard the jelly or aspic as essential; it really sees macédoine of fruit as what might otherwise be called simply fruit salad, while macédoine of vegetables is a mixture of finely diced carrots, turnips and beans, with peas and asparagus tips, which can be served hot or cold. The French also use the word macédoine for a book put together out of snippets from a variety of sources, a hotch-potch. And the origin of the word? Littré, the greatest of French dictionary makers, is cagey – first wondering whether there is not some specific allusion which time has buried, before tentatively suggesting that there may be a reference here to the empire of Alexander the Great, who set out from Macedonia to conquer a vast empire that was made up of a great variety of different states and kingdoms.

Madras

As well as producing Madras net – a fairly coarse muslin – and Madras handkerchiefs, which were much appreciated by the blacks in the West Indies on account of their bright colours, this city on the south-east coast of India is famous in educational history for the so-called Madras system. This was devised by Andrew Bell, a Scottish clergyman who found when he tried to organize schooling for the inmates of the boys' orphanage in Madras in 1789, that progress was difficult because there were not enough teachers available. So he had the idea of arranging classes in such a way that the master taught a number of the more able and advanced pupils, who then passed on to the others what they had learned. He found that results were good and decided that the system could be applied generally. On returning to Britain in 1797 he published a booklet describing his methods, but little heed was paid to them until they were taken up by Joseph Lancaster, a Quaker who set up a school on similar lines in Southwark, which led to the name Lancaster system (or more generally monitorial system). It was mainly because members of the Church of England were so alarmed to see educational reform becoming a nonconformist preserve that Bell was called to London from a country rectory to found the National Society for Promoting the Education of the Poor in the Principles of the Established Church. On his death Bell left considerable sums for educational purposes and a school called the Madras Academy was founded in Cupar, Fife.

Magenta

Magenta is a town in Lombardy where an army of Piedmontese and French soldiers under Napoleon III defeated the Austrians on 4 June 1859. The victory is commemorated in the name given to a brilliant crimson aniline dye discovered shortly afterwards. The other French triumph of that campaign was at Solferino on 24 June 1859. That battle too is commemorated in a dye, Solferino red, which has a bluish tinge. A more interesting fact is that the carnage was so terrible that Henri Dunant was inspired to found the Red Cross.

Mafficking

As every Boy Scout knows, people went mafficking in May 1900 to celebrate the relief of Mafeking, a township in what was then called British Bechuanaland rather more than 150 miles west of Johannesburg. The Boers besieged the place at the outbreak of hostilities in 1899, but the British garrison led by Colonel R. S. S. Baden-Powell – who showed genius in keeping up morale against all odds – held out for 217 days. When news of the arrival of a relief column reached London, the crowds went delirious. That was perhaps understandable, because there had not been a lot to crow about during the Boer War until then and, truth to tell, there was not to be much to celebrate in the concluding months of the conflict either. Partridge notes that the term was revived in November 1918 and May 1945 and comments that 'apparently it is here to stay'. That appears doubtful on linguistic grounds alone, apart from any thought of political or military considerations. Because 'mafekinging' would be impossibly difficult to manage, especially after a few pints, the shortened form 'mafficking' had to be coined. But that produces the verb 'to maffick' and then all immediate meaning drops out. On balance it seems likely that if the word ever reappears in journalism, it will be a conscious – and not very comprehensible – echo of imperial jingoism.

Maltese cross

A Maltese cross has four arms which expand in width as they lead out from the centre. It is the badge of the Knights of St John of Jerusalem to whom the Emperor Charles V granted the island in

1530, and who had to make good their right to stay there when the Turks laid siege to it thirty-five years later. The Knights maintained their independence until 1798 when they surrendered to the French, and the island came formally into British possession in 1814, attaining self-government in 1947. Before then, in 1943, in commemoration of the islanders' steadiness under bombardment in the Second World War, Malta had been awarded the George Cross. In legitimate pride this is depicted in the national flag, but the irony is that the George Cross is of the Greek and not the Maltese pattern. The Maltese cross is, however, emblazoned on the island's maritime flag. Because of the pattern of its brilliant red petals, the Scarlet lychnis (*Lychnis chalcedonia*), which thrives in boggy situations, is commonly called the Maltese Cross.

Manila envelopes

In theory Manila envelopes are made with paper produced entirely from manila hemp fibre, which is the longest and strongest fibre employed in paper-making. Manila paper, which in fact is generally made with wood pulp, is typically machine-glazed on one side and left rough on the other. Manila is the capital of the Philippine Islands; its port was opened to foreign trade in 1837 and on 1 May 1898 a United States squadron under Commodore Dewey destroyed the Spanish fleet there.

Manhattan

Enough of dry etymologies! Let's just fix a Manhattan this time and relax.

1 jigger of rye whisky
1 jigger of Italian Vermouth
dash of aromatic bitters

Stir well with cracked ice, strain into a cocktail glass and decorate with a maraschino cherry. A jigger, in case you have forgotten, is 1½ fluid ounces.

Marathon

Marathon comes from a word meaning 'fennel' which gave its name to the plain by the sea where the Greeks under Miltiades inflicted a signal defeat on the Persians in 490 BC. Pheidippides ran to Athens to give the good news and he covered the whole distance of approximately 22 miles without stopping. A marathon race was included in the first modern Olympic Games at Athens in 1896, but it was not until the 1908 London Olypmics that the distance was standardized. The runners had to cover not only the 26 miles from Windsor to the White City Stadium, but an extra 385 yards so that the finishing line was opposite the royal box. 'Marathon' has come to mean anything long and demanding; with the record for the marathon at just over two hours, it looks as though we may be needing a new adjective before long.

Marcella

Marcella is a type of cloth associated with the great French seaport of Marseilles. Formerly in demand for fancy waistcoats, it is a fabric of some thickness made from cotton or linen, nowadays used for the textured fronts of dress shirts.

Marseilles also gives its name to the French national anthem, though the story of how it did so is rather complicated. On 25 April 1792 the mayor of Strasbourg was entertaining a number of army officers when news arrived that the revolutionary French Republic had declared war on Austria, and he remarked how unfortunate it was that the troops lacked a stirring battle hymn. Claude-Joseph Rouget de Lisle, a distinctly minor poet who was serving as an engineer, accepted the challenge and it is said that he composed both the words and the tune of what subsequently became known as the *Marseillaise* that very evening. He himself gave it the title 'The Song of the Army of the Rhine', but it was taken up by patriots of Marseilles who sang it when they came to Paris and showed their solidarity with the Revolution. It was sung by the army by official order at the Battle of Valmy on 20 September 1792. Banned by Napoleon in a reactionary moment and by the Restoration, the *Marseillaise* has enshrined a characteristic aspect of French patriotism. King Louis Philippe, whose policy was to reconcile conflicting pressures in France after 1830, gave the aging Rouget de Lisle a pension and in 1915, in a display of what might be described as secular hagiolatry in France's hour of desperate need, his ashes were solemnly transported to the Pantheon in Paris. There is a setting of the *Marseillaise* by Hector Berlioz which captures all its fervour.

Maris Piper

The success of Maris Piper qualifies it for a mention here, and the entry offers an opportunity to point out that EEC regulations are now concerning themselves with the suppression of what many might have supposed harmless toponyms. The Maris Piper is a main-crop potato which has contributed in a large way to the dethroning of Majestic. It comes from the Plant Breeding Institute at Trumpington, near Cambridge, which in the sixties made a practice of combining a reference to its address in Maris Lane with some other suitable term in the name of new varieties developed there. Purist lexicographers might protest that Maris Lane was in its turn derived from the name of the Maris family who once lived there, but we can be reasonably confident that the plant breeders were not out to commemorate former occupants of the site. Other famous varieties from the Institute include Maris Huntsman, a famous high-yielding wheat, and Avalon, which has become a very widely grown winter variety.

Marsala

But for John Woodhouse, a Liverpool merchant who visited Marsala – a seaport on the extreme eastern end of Sicily – around 1760, marsala might never have become the wine (or the toponym) that it is. It is a fortified wine prepared in a complicated fashion by adding not only brandy but also a sweeter wine and treated unfermented grape juice to the original white wine of the district. The result is a drink of some character which is, to my mind at any rate, no better when flavoured with strawberry. Apart from being an interesting dessert wine, marsala is the essential ingredient of zabaglione. Take:

3 egg yolks
4 dessertspoonsful of sugar
¼ bottle marsala

Whisk energetically and then cook in a microwave on high for 5 minutes, removing from oven once every minute and whisking up again until frothy. Finally whisk once more and pour into glasses, decorating with stiff cream. Try it – it's a better use of the microwave than those endless baked potatoes.

Martello

Martello towers take their name from Cape Mortella in Corsica, where in 1794 a low round artillery fort with a small garrison and just three guns proved its worth when attacked from land and sea by strong

British forces which had come with the intention of supporting Corsican insurgents against the French. The experience convinced the British authorities, and during the invasion scare in the course of the Napoleonic wars large numbers of Martello towers were built around the south and east coast of England. Whether they would have stopped Napoleon is a moot point, but many of them have survived to become seaside museums and the Martello tower at Aldeburgh – allegedly built at Nelson's suggestion as protection to the river at Slaughden Quay – has been turned into a holiday home which you can rent by the week from the Landmark Trust. The pepper-pot towers which are a feature of Guernsey beach scenery, especially at L'Ancresse Common, should not really be called Martello towers; they were built before the British had so much difficulty at Cape Mortella and belong to an earlier, less scientific tradition of military engineering.

Mayonnaise

Mayonnaise is the second half of the French culinary term *sauce mayonnaise*, which was coined around 1756 to commemorate the capture by the Duke de Richelieu of the anchorage of Port-Mahon and of Minorca, in the Balearic Islands, from the British who had been there since 1708. The original formulation, *mahonnaise*, was phonetically very unstable and soon displaced by something easier to say in French, even if the history was half-hidden.

Mazurka

The polonaise is, obviously enough, Polish in origin; it began as a kind of triumphal dance performed to celebrate the return home of a victorious monarch. In triple time, it became especially popular for the opening of balls, when it would be led by a couple who indicated what patterns the rest of the company should adopt. Interpreted for the piano with incomparable verve by Frederic Chopin, the polonaise breathes the spirit of the nation. He also gave us his versions of the mazurka, another Polish folk dance, also in triple time; this originated from the province of Mazovia and came into European ballrooms in the second half of the nineteenth century. Polka sounds as if it ought to be a Polish dance too, but in fact this came from Bohemia.

Meander

The Meander was a river which twisted and turned as it crossed the Phrygian Plain outside Troy on its course towards the sea. The word has left a rich deposit in usage. Physical geographers talk of the meanders of rivers and poets sing of streams that meander along. For art historians who have forgotten their Homer for a moment, a meander is an ornamental pattern in which the lines wander and criss-cross at right-angles. And we all complain about political speeches which meander endlessly on and never arrive at any conclusion.

Meringue

Meiringen is a small town in Switzerland – in Haslital, Unterwalden canton, east of Interlaken – which claims to be the home of the meringue. The story goes that a local chef, one Gasparini, whipped up left-over egg whites to create the cakes to which none other than Napoleon is supposed to have given the name of 'meringue'. A report in the *Daily Telegraph* for 9 February 1985 tells how all hands in Meiringen are united in an effort to win a place in *The Guinness Book of Records* by making the world's largest meringue. Good for them! The only problem is that the *O.E.D.*, while cautious about proposing an etymology, cites examples of the use of the word 'meringue' in English as early as 1706, whereas Napoleon was not even born until 1769. Whatever the truth about origins, it appears that the Meiringers have taken meringues to their hearts. Tourists can always eat them there when in the region to make a pilgrimage to the Reichenbach Falls where, if Conan Doyle had had his way, Sherlock Holmes would have drawn a last watery breath.

Millinery

Originally milliners were people who sold all kinds of cloth, haberdashery and fancy goods, and it was only in the last century that the term came to be particularly associated with women who sold ladies' hats. The word 'milliner' is derived from 'Milaner', in other words a person who lives in or comes from the Italian city of Milan. It was used in that precise sense in the early sixteenth century, but the transferred meaning was already current by then too. At that time the name of the city was stressed on the first syllable (Mílan, not Milán) which meant that the word fitted in with normal English rhythms. The ending –ery of millinery served to relate the word to grocery, grindery and other similar terms.

Mint

Mints, and in fact the word money and its derivatives, come from the temple of Juno Moneta on the Capitoline Hill where the Romans set up their first factory for the production of coins. Whether Juno Moneta means 'Juno of the Mint' or 'Juno, goddess of counsel' is a matter of some dispute, but about the link with the temple there can be no doubt. After being housed for five centuries in the

Tower of London, the Royal Mint moved into premises on Tower Hill early in the nineteenth century. In 1968 the Queen opened the first phase of the new Royal Mint at Llantrisant in South Wales.

Mocha

Mocha, properly the name of a particular kind of Arabian coffee and more generally a genteel adjective meaning simply 'coffee-flavoured', comes from Mokha which is a town on the Red Sea coast of Yemen. At one time it was the chief port of the export of coffee grown in the mountainous hinterland. Mocha stones, a variety of chalcedony with markings more or less like trees or branches, used to be exported through Mokha too, though nowadays they are obtained from India.

Montelimar

Montélimar is a small town in south-east France, in the department of Drôme, on the left bank of the Rhône about a hundred miles south of Lyons. It has some fame as a centre for the manufacture of nougat, and a chocolate-coated variety figures in the more expensive boxes of sweets.

Morris dancing

It seems fairly certain that the English folk dance derived from the Moresca, which was known in Burgundy from the early fifteenth century. One idea is that there might be a connection with words like 'moron', suggesting that the dance represented in some degree the age-old folly of carnival. A more likely explanation however is that the link is with the Spanish *morisco*, which takes us back to the time when the Moors were in Spain.

Muslin

Muslin is supposed to have been made first at Mosul, a city in Mesopotamia. The diminutive suffix *-in* was perhaps a tribute to its fineness; oddly enough 'muslinette' was not finer still, but thicker, though the cloth like the linguistic impropriety seems to have disappeared from use. The art of weaving muslins spread from the Middle East to India and it was from there that they were first exported to Britain in the seventeenth century.

Newmarket

Newmarket, though only a small town, has been famous for its associations with horse racing – and hence with everything that is sporting and fashionable – since the time of James I. There is a card game called Newmarket, which in America is known as Michigan. A Newmarket jacket was a riding coat cut away at the front and with full skirts to the back; in various forms including ladies' models, it was worn throughout the nineteenth century.

Norfolk jacket

If you don't know what a Norfolk jacket looks like, the easiest way to find out is by looking carefully next time you see a photograph of the Duke of Edinburgh when he is properly dressed for leisure in the country.

Norfolk jackets, which are made of tweed, button fairly high in front and have pleated patch pockets, but the distinguishing feature is the belt made of the same material which is stitched to the jacket at the back. Introduced in about 1876, when they were described as suitable for any kind of outdoor exercise, by the 1890s they had been adapted by young women with advanced opinions and a taste for ladylike sport. For some reason Norfolk jackets have never really caught on with the general public, and now competition from the anorak and the parka will probably stand in their way. But for anyone who can afford the fairly steep initial price and who is not afraid of putting on too big a paunch, a Norfolk jacket will remain a sound investment.

Orpington

Orpington in Kent, famed in the annals of the Liberal party for a by-election sensation, has also given us some poultry. There is the Orpington chicken and her antipodean cousin, the Australorps, both notably large birds which lay brown eggs. The Buff Orpington is a duck to tickle a toponymist's fancy; it was bred by crossing the Aylesbury, the Rouen and the Indian Runner from Malaya.

Osborne biscuits

Plain, sweet and not especially distinguished, Osborne biscuits were named after Queen Victoria's favourite residence at Osborne on the Isle of Wight. Whether she was amused is not recorded, but her loyal subjects no doubt approved of the implied connection. Queen Victoria purchased the estate which lies south-east of Cowes in 1845, and she and Prince Albert took great personal interest in the building here of a Palladian villa. It was there that the Queen died in 1901, supported on the arm of Kaiser Wilhelm. Edward VII, who had made a country home for himself at Sandringham, quickly disposed of Osborne, which was then used as a convalescent home for Navy and Army officers. In

1903 a Royal Naval College was opened in premises on the estate.

Opus Anglicanum

Opus Anglicanum is an aristocrat among toponyms and the honorifics of a Latin title are not unmerited. English embroidery was noted from Saxon times onwards, but it was in the thirteenth and the first half of the fourteenth century that English ecclesiastical embroidery reached its peak. It was exported to the Continent, and it may be that the need to satisfy demand led eventually to some decline in standards.

Oxford

As well as lending its support to lost causes, Oxford has given us an intriguingly wide range of toponyms. The serious side of the university is commemorated in the Oxford Movement, which gingered up the Church of England in the last century. Fashion is represented in Oxford shirting, which was striped; Oxford bags, which were extremely wide trousers much affected in the years following the First World War; and Oxford shoes, which have the front edge of their lace-eyelets stitched neatly into the vamp (unlike Derbys – named after the earl – which leave them as flaps). Oxford corners are found in printing, where the text is enclosed within ruled lines which extend to form a cross in each corner of the page. An Oxford accent seems to be recognized the whole world over, except in Oxford itself, which is really very puzzling.

Pacifics

Pacifics are among the largest steam locomotives to come into general service. For railway buffs, it is enough to say 4-6-2, which means that between two pairs of smaller wheels at the front and one pair at the rear there are no fewer than three pairs of great driving wheels. Pacifics were introduced by the Pennsylvania Railway in 1905; they marked a distinct step up from the Atlantics – 4-4-2 – which had appeared on the Atlantic Coast Line in 1894, and the name marks their superiority over these without however having any specific connection with the other Ocean. They soon established themselves in the USA, and in 1922 Sir Nigel Gresley designed Pacifics for the Great Northern Railway. The locomotive also occupies a corner of musical history, because the Swiss musician Arthur Honegger composed a tone poem called 'Pacific 231' in 1924, in which he expressed its power and rhythmic vitality.

Paisley patterns

Paisley in Renfrewshire, Scotland, was so famous in the nineteenth century for its silk gauzes and fine shawls that the characteristic design based on the shape of the pine cone came to be known as 'Paisley pattern'. The motif had in fact originated in India and had been used in many centres on the continent and in England before it was taken up by Scottish manufacturers.

Palace

Rome was built on seven hills, one of which was called the Palatine Hill after Pales, a pastoral deity. It was here that Augustus built his grand residence, and Tiberius and Nero followed his example. In time this gave us the word 'palace'. Originally built of red brick for the Duke of Buckingham in 1703, Buckingham Palace was remodelled by Nash in 1825 and took on its present appearance in 1913, when Sir Aston Webb gave it a new façade of Portland stone.

Panama hats

Gentlemen's lightweight white straw hats for summer use were widely known as Panama hats, even though they might have been manufactured in Luton, which was a centre of the trade. Properly speaking, Panama hats were made of the leaves of the screw pine, *Carludovica palmata*, which were bleached before being plaited. However, even they had no particular connection with Panama itself.

Pantechnicon

Probably there is something apt in the lumbering polysyllables of this word freighted with Greek which has made it stick as the name for large removal vans. The etymology takes us to 'all the arts', but that hardly hints at the toponym behind the term. Around 1830 a bazaar selling goods of every description was opened in Motcomb Street, in London's Belgravia, and it was called the Pantechnicon. After a while this became a furniture warehouse and the vans which served it took their name from it.

Paramatta

Paramatta is an exotic among toponyms and seems to have been used only because the sound of the name appealed to somebody's romantic imagination. Parramatta, with two r's, is a town in New South Wales, a few miles north-west of Sydney. Founded in 1788, it was one of the earliest inland settlements in those parts and became an important administrative centre. Tweed was manufactured there, but it is very doubtful whether that has anything to do with the dress cloth made with a woollen weft and a silk or cotton warp, known to the Victorians as paramatta.

Parchment

For centuries people have written the records that they are especially anxious to preserve on a material made from the treated skins of a sheep or goats – to which is given the name of 'parchment'. This is rather apt, for the word commemorates a city famous in antiquity. Borrowed from French which had adapted it from Latin, the noun comes from Pergamon, a city in Mysia on the coast of Asia Minor which was famous for its culture.

Paris Green Bait

Paris Green Bait is used to control earwigs and other leaf-eating plants. A copper aceto-arsenite compound, it was formerly used extensively in horticulture.

Peaches

Peaches were probably first grown in China, but it was from Persia that they were imported into Europe. The Romans simply called peaches 'Persians', leaving out the unnecessary noun. English picked up the word in its Frenchified form and started using it in Chaucer's time. The scientific name for the peach tree is still *Prunus* or *Amygdalus persica*, i.e. Persian plum or almond.

Pekingese

Pekingese dogs have been carried here – for it is a fair bet that they could never have walked the distance – from Pekin, which seems to be spelt a different way every time our relations with the Chinese People's Republic get a little better. Only the Emperor of China and his immediate circle were officially allowed to keep the so-called palace dogs, whose build and features were supposed to have symbolic significance. They were not known in the West until they were brought back after the looting of the Imperial Summer Palace at Pekin in 1860 – an act, as Lytton Strachey remarked with icy disdain, 'by which Lord Elgin, in the name of European civilisation, took vengeance upon the barbarism of the East'.

Pembroke tables

These are the convenient and attractive tables which have four fixed legs to support the top and two flaps which can be raised and then supported on hinged brackets. Why they have the name is uncertain. Some authorities suggest they are called after the Welsh town. Others, with perhaps rather more justification, quote the cabinet-maker Thomas Sheraton's comment that they were named after the lady who first had the idea for the design. If that is correct, he must have been referring to the Countess of Pembroke (1737–1831); the last two-thirds of her life cover the period when Pembroke tables were most fashionable.

Pentland Crown

Pentland Crown has been a highly successful main-crop potato. Like Pentland Javelin, an early variety which achieved an almost equally spectacular success and quite eclipsed the once very popular Arran Pilot, it was bred by the Scottish Plant Breeding Station which in 1954 moved to a site near Roslin, Edinburgh, close to the Pentland Hills. For some years thereafter, varieties produced by the Station were given the prefix 'Pentland'. After 1970 two-part names of this sort were forbidden by EEC legislation, so there has been a switch to names with a Scottish flavour to them such as Provost, Baillie, Moira and Sheena.

Pheasants

In what was anciently called Colchis, the region at the far eastern end of the Black Sea, is the River Phasis (now Rioni). Legend has it that the Argonauts under Jason brought back from those parts not only the Golden Fleece, but also the birds which ever since have been known in European languages by one form or other of the word 'pheasant'.

Piltdown Man

Toponyms serve to chart some of the more interesting stages in human palaeontology as mankind tries to piece together his ancestry from cracked skulls and broken bones found scattered over the earth. Discovered in the 1920s in a cave in Dragon Hill at Choukoutien, south-west of the Chinese capital, Peking Man walked upright 400,000 years ago. Neanderthal Man, famous for his massive brow, is called after the valley of the River Neander in Germany, a few miles east of Dusseldorf, while the Old Man of Crô-Magnon takes his name from Crô-Magnon, near Les Eyzies in the Dordogne region of France.

The joker in the pack is Piltdown Man, an elaborate fake purportedly discovered by the amateur archaeologist Charles Dawson in a gravel pit in Sussex, not far from Uckfield, in 1912. Despite some scepticism, the scientific establishment was generally taken in and not until 1955 was it established that the find had in fact been produced by combining a human cranium and an orang-utan's jaw. The episode is commemorated in a pub which bears the name 'The Piltdown Men' and also in a number of red faces. The toponymous tradition is continued in this sphere by Selsdon Man, a term applied to Tory policy-makers who gathered in the Selsdon Park Hotel near Croydon to make plans before the Heath government was elected.

Pistol

In Tuscany twenty miles north-west of Florence is a town called Pistoia, where the short dagger called the 'pistolese' was manufactured in the late Middle Ages. The name was transferred to small arquebuses; then it acquired the diminutive suffix 'et', to become yet smaller. Taking over the word from the French, the English docked a final syllable that served no useful function.

Plaster of Paris

Plaster of Paris is fine white anhydrous gypsum which swells when wetted and then quickly sets hard. It is used particularly by modellers and decorators and for making plaster casts for broken bones. Gypsum for plaster of Paris came originally from Montmartre, the hill now crowned with the basilica of the Sacré Coeur which dominates the northern skyline of the capital. There have been suggestions that Montmartre itself was first called *Mons Mercurii* because of the Roman temple dedicated to Mercury that once stood there. It seems far more likely however that the present name means 'Mountain of the Martyrs', in commemoration of the execution of St Denis there in the year 272.

Plymouth Brethren

Plymouth has given us a great deal. There are those stirring tales of Sir Francis Drake finishing his game of bowls on The Hoe before sailing off to give the

Spanish Armada a drubbing. Then the *Mayflower* changed another chapter of history, and more recently the city has distinguished itself by an enterprising choice of Members of Parliament – Nancy Astor, Dame Irene Ward and Dr David Owen.

The crop of toponyms is interesting too. A 'Plymouth cloak' is old slang for a cudgel carried for self-protection. Plymouth gin, as the recipe provided under 'Angostura bitters' shows, is an essential ingredient for a genuine pink gin. And, in more reflective vein, there are the Plymouth Brethren. The Reverend John Nelson Darby (1800–1882) was an Irishman who had been ordained into the ministry of the Church of Ireland, but renounced his orders because his reading of the Scriptures made him doubt the validity of existing religious institutions. In 1830 he preached in Plymouth and a number of people declared themselves for his principles. Features of the Plymouth Brethren beliefs are devotion to the Bible, especially the New Testament, and insistence on the autonomy of each local church. There is a strong element of Calvinism in their beliefs, and they will undertake only those occupations which they regard as entirely compatible with the teaching of the New Testament. Though never attracting very large numbers of converts, this religious community has spread far and wide, both in Europe and further afield, and at least six sub-groups have been distinguished amongst what can generally be called the Plymouth Brethren. Plymouth Rocks are a heavy breed of poultry which produces tinted eggs; varieties include Barred Plymouth Rocks which have white plumage with heavy black bars that have a green sheen, and White Plymouth Rocks which are bred mainly for the table.

Polony

Polony, the sausage made partly with uncooked pork, does not come from Poland. As early as the mid-seventeenth century the fairly common phonetic shift had occurred, and Polony and Polonian had emerged from Bologna or Bolognian sausage which was associated with Bologna, in northern Italy. The same city has of course also nourished our vocabulary with spaghetti bolognese.

'Baloney', meaning nonsense, came from America in the twenties. There is some doubt about the correctness of relating it to the Italian sausage in whatever phonetic skin, but etymological scruples are no threat to such an expressive onomatopoeia.

Pontefract cakes

Pontefract in Yorkshire is supposed to be pronounced Pomfret, and the name seems to refer obviously enough to the breaking of the bridge there, though why and exactly when – at some date between the Norman Conquest and the mid-twelfth century – remains a subject for conjecture. As for the black liquorice sweets the size and shape of a tenpenny piece, they trace their origins back to the times when Cluniac monks brought back liquorice bushes from the Mediterranean. Originally used purely for medicinal purposes, the liquorice began to be used for confectionery towards the end of the eighteenth century. The firm of Wilkinsons, that still makes Pontefract cakes, celebrated its centenary in 1984.

There are of course many other towns which boast confectionery specialists, even if they sometimes amount to little more than a familiar product under a suitably allusive name. Berwick Cockles, for instance, are boiled sweets from Berwick-upon-Tweed. Edinburgh has its own variety of rock which, unlike the one upon which the Castle stands to dominate the city, is soft and soon dissolved, while every seaside resort can rise to some sort of lettered rock, with Graham Greene investing Brighton Rock with fresh dimensions. Harrogate, in the guise of John Farrah whose shop opened in 1840, makes a bid for pre-eminence among Yorkshire toffees. From a delightful Westmorland town, Kendal Mint Cake has always sought fresh peaks to conquer and the wrappings are eloquent accounts of a few moments' satisfied munching high up in the Himalayas.

Port

Port is a thoroughly English toponym, and to make it feel quite at home here, the 'o' has been chopped off from each end of the original place name whence it is derived – Oporto. British merchants had been settled in Portugal since the sixteenth century, but the wine trade did not really develop until after 1703 when the Methuen Treaty – designed to damage the French economy – allowed the import of Portuguese wines into England at advantageous rates. Since then, despite changes in excise duties, port has hardly looked back; it is produced from wine made from grapes grown in the Douro region of Portugal and exported through the harbour of Oporto on the Atlantic coast.

Port Salut

Port Salut – or better, Port du Salut – is the soft cheese which was the creation of Trappist monks (q.v.) who were banished from France during the Revolution and settled in Switzerland. When they were allowed back to their homeland, they returned to the Abbey of Notre-Dame de Port du Salut (Our Lady of Safe Refuge) in the parish of Entrammes near Laval. The monks were selling the cheese by 1850, and when it went on to the market commercially in Paris in 1873, it was a great success.

Pomeranian

Two questions arise. Where is Pomerania? And have Pomeranians anything essential to do with it? Formerly Pomerania was a province of Prussia, on the Baltic coast east of Mecklenburg. Now the territory east of the Oder, including the city of Szczecin (Stettin), has been ceded to Poland. The story of the breed is that around 1800 a number of German spitz dogs were brought to England, where at first they were called Pomeranians. Quite often tiny puppies were found in normal litters, but it was not until the end of the nineteenth century that any interest was taken in raising a breed from these apparently undersized specimens. The result was the miniature variety, called the Toy Pomeranian, which had not previously been known in Germany.

Portguese men-of-war

Every now and again Portuguese men-of-war are reported in the Channel, but that is never an excuse for lighting signal beacons to announce that a fresh armada is approaching our shores. Portuguese men-of-war – *Physalia* to marine biologists – are animals related to the corals and the jelly-fish that float on the surface of the sea. With their brilliant blue bladders with a crest on top, they sail along as stately as caravels. But though they look harmless, they have an underwater weapon – retractile tentacles sometimes a couple of yards long with batteries of tiny capsules that can sting as sharply as nettles.

Prussic acid

As well as Prussian blue – the deep blue pigment discovered by Diesback in 1704 – the German state and cyanide compounds combined to give us prussic acid, the virulent toxin beloved of Victorian poisoners

Quince

We shall have to peel this word to get back to the place where it originated. Centuries back 'quince' was the plural of 'quine', which itself had evolved from 'coin' (close to the French form *coing*, in which the *g* is not pronounced). 'Coin' in its turn comes from Cydonia, or rather Pyrus Cydonia, meaning the pear tree from Cydon, a city in Crete. A related genus is the japonica, of which the three species occur naturally in Japan and China.

Russia duck

'Some were, as usual, in snow-white smock-frocks of Russia duck.' This quotation from *Far from the Madding Crowd* is indication enough that 'Russia duck' was not a bird but a fabric, in fact a strong linen. A Muscovy duck, on the other hand, is a bird, but it has no connection with Russia, the adjective being a corruption of musk and the species coming originally from South America. The low quality unrefined sugar known as muscovado sugar has nothing to do with Russia either; the word is from the Spanish. Russia leather, however, is noted for its strength and is used for book-binding; it is prepared by tanning in oil distilled out of birch bark.

Roman candles

Roman candles are fireworks made by filling cylindrical cases with alternate layers of explosive compound and stars made out of filings of various metals so that colourful effects are produced as they fall after being tossed high into the air. How they came by their name is unclear; it hardly seems likely this can be an allusion to the rhyme beloved of schoolboys:

> The Emperor Nero
> Was no hero;
> He used Christian communicants
> As his illuminants.

Bengal lights are fireworks too, designed to produce beautiful effects; originally they gave off a blue light which was used for signalling.

Rugger

Rugger is the toffee-nosed corruption of Rugby football which, in some fanatical areas, has achieved the supreme status of being called simply football. The origins of football are buried in the mud and gore of centuries of popular rivalry between villages. What distinguishes Rugby from other varieties of the game is summed up in a memorial plaque set in the wall of the Close at Rugby School, the famous public school founded in 1567, reformed by the redoubtable Dr Thomas Arnold and glorified in *Tom Brown's Schooldays*. The inscription runs: 'This stone commemorates the exploit of William Webb Ellis, who, with a fine disregard for the rules of football as played in his time, first took the ball in his arms and ran with it, thus originating the distinctive feature of the Rugby game. AD 1823.' As regards the game, that is quite illuminating, but doubts have been cast on the accuracy both of the date and of the exact role played by Ellis. Be this as it may, the connection with Rugby appears certain and the name has stuck. 'Soccer', from Association Football by analogy with 'rugger', has been more widely accepted as a term than 'rugger', probably because the latter was too obviously linked with public school speech habits to come readily to those who play the game in less rarefied circles and to their professional confrères who play the thirteen-a-side variant known as Rugby League.

Ruritanian

Ruritania, like Utopia, does not exist. The figment of the fertile imagination of Anthony Hope (the diplomat Sir Anthony Hope Hawkings), it was the deliciously bogus quasi-Balkan setting for Rudolf Rassendyll's adventures in *The Prisoner of Zenda* (1894) and *Rupert of Hentzau* (1898). People persist in using the term 'ruritanian' when comparing this or that event in real life, but truth will never quite live up to fiction in this case.

St Bernards

Lexicographers, as Dr Johnson almost said, have a dog's life and it would be possible to split hairs deciding whether St Bernards are eponyms or toponyms. The likelihood is that they belong to the latter category, being named after a place which had previously been called after a person.

The Hospice of the Great St Bernard Pass between Switzerland and Italy stands more than 8000 ft above sea level in the canton of Valais; it was originally built under the direction of Bernard de Menthon, who was subsequently canonized in the year 962. It is not certain when dogs were first employed there to help the monks in their work of succouring travellers in distress; this may not have been until the middle of the seventeenth century.

Nineteenth-century sensibility, which found tales of animals' devotion to the human race in times of difficulty very affecting, made a lot of the St Bernard; and Longfellow summed it all up in 'Excelsior!', the tale of the youth who struggled up the Alpine slope and came heroically to grief:

> A traveller by the faithful hound,
> Half buried in the snow was found,
> Still grasping in his hand of ice
> That banner with a strange device,
> Excelsior!

According to the official description, the expression of the St Bernard should betoken benevolence, dignity and intelligence.

Sandwich tern

The Sandwich tern is indeed named after the little Cinque Port in Kent where its presence was first recorded in England. The Latin name is *Sterna sandvicensis*; the adjective is unremarkable, but the noun *Sterna* is a reminder that in Norfolk terns were called starns until the nineteenth century. Sandwich terns come to Britain from the Mediterranean to breed in coastal colonies as far north as the Orkneys. The other words with Sandwich in them, including a number of islands, relate to the fourth Earl of Sandwich, John Montagu (1718–1792). After being educated at Eton and Trinity College, Cambridge, he achieved notoriety for corruption and incapacity while First Lord of the Admiralty and won fame for surviving for twenty-four hours at the gaming table with no sustenance other than beef sandwiches which we have called after him.

Sardine

The larger dictionaries are surprisingly hesitant about connecting sardines with the Mediterranean island of Sardinia (which is itself so-called because its shape was supposed to resemble a footprint). Sardines are found there in large shoals, however, and there seems to be no better etymology available for the name given to these small fish belonging to the herring family which are pickled in brine or oil and sold tightly packed in jars or tins. It is of course the packing rather than the fish which next enriched social life with the game of sardines. Not so much a parlour game as an escape-from-the-parlour game, its rules are simplicity itself. One member of the party is sent out to find a hiding place where he or she can lie low. After a reasonable delay the remainder of the group set out in search, but if they find the hider they do not inform anyone else; instead they simply join him or her and wait, amusing themselves as best they can in total silence until they are joined one by one by the rest of the party. It can take hours, with a bit of luck, and generally nobody is concerned to charge the least successful searcher a forfeit. But things can go wrong! In the ballad 'The Mistletoe Bough' by that expert purveyor of hokum Thomas H. Bayly, Lovell's bright-eyed bride hid away in an old oak chest; the lid fell down and snapped shut and she was not found again for years.

Satsuma

It was around 1700 that the mandarin was brought from China to Japan, where it did especially well in the southern province of Satsuma, on the island of Kiusiu.

Saunter

Not entirely inappropriately, etymologists are by no means certain where the verb to saunter came from. An agreeable if not very scientific suggestion is that it might be connected with 'Sainte Terre', the Holy Land towards which pilgrims often seemed to be making their way with no great sign of haste.

Savonnerie carpets

In French 'savonnerie' means soap-works, and a little history will suffice to explain why savonnerie carpets came to set new standards in Europe. It was Pierre Dupont who was the first Frenchman to imitate successfully the Turkish knotted-pile of carpet making. As a sign of royal favour, Henri IV arranged for him to have premises in the Louvre Palace, Paris, in 1606. Twenty years later, when the old soap-works on the Quai de Chaillot – near the present-day Place d'Alma – fell vacant, it was acquired in order that production could be increased. Run at first under the direction of Simon Lourdet, Dupont's partner, the work-force was recruited largely from orphans who were apprenticed to the craft. Though carpets continued to be made at the Louvre for another half-century, the use of the name 'savonnerie' became generally current. It was not until 1768 that savonnerie carpets became available to the public, and even then prices were dauntingly high and little was offered for sale. Napoleon, with his passion for imperial splendour in his palaces, naturally made heavy demands on the Savonnerie. In 1825 it was amalgamated with the Gobelins works, where some of the original carpet looms are still used.

Savoyard

Savoyards – originally those who took part in the first performances of the operettas – are now the knowledgeable fans of Gilbert and Sullivan. The name derives from the Savoy Theatre in London which Richard D'Oyly Carte opened on 10 October 1881, transferring *Patience* there from the Opéra Comique where it had been running since the spring. The new theatre was designed with great attention to the comfort of the public; more significantly still in the history of theatre, it had the distinction of being the first to be lit throughout by electricity. The interior was, however, completely replaced in 1929. The Savoy Theatre stood near the site of the former Savoy Palace, erected in 1245 by Peter of Savoy who was invited to come to England from Italy by Henry III and then made Earl of Richmond. The property passed into the hands of John of Gaunt, Duke of Lancaster, and though the palace was destroyed in Wat Tyler's rebellion in 1381, the Queen's Chapel of the Savoy remains the personal property of the monarch.

Scarborough warning

Giving 'Scarborough warning' of doing something means doing it unexpectedly, allowing not a moment for taking any precaution against the action. It is an old expression whose origins are doubtful, though it appears a reasonable surmise that it refers to some maritime incident when ships made a sudden attack. It shows how incalculable are the ways of idioms that 'Scarborough warning' was not given a fresh lease of life after German battle cruisers came out of the North Sea mists at dawn on 16 December 1914 and shelled the town for a while before scampering back to their base. Scarborough also suffered bombardment from submarines in April and September 1917.

Scarborough Ulsters, first seen in 1892, were long overcoats of heavy tweed with cape and hood, but no sleeves; Scarborough waterproofs made their appearance four years later. There is also a plant called the Scarborough lily, though whether it has any connection with the resort seems doubtful. Otherwise known as the Vallota, it is a bulb genus from South Africa with only one species that grows up to a height of two feet, and its flowers are bright scarlet. The Scarborough lily grows outdoors in the UK only in the far south-west.

Scarper

'Scarper', slang for running off or leaving without due notice or paying the bill, is recorded as early as the mid-nineteenth century and its origins are probably Italian. Around 1918, though, the variant spelling 'scarpa' started to crop up; it is probably Cockney rhyming slang, linking 'go' with 'Scapa Flow', then in the news as the Grand Fleet's base in Orkney during the First World War. It is tempting to speculate on some further reference to the incident when the German warships interned there under the Armistice scuttled themselves on 21 June 1919. There may be something in the idea, but the evidence will hardly stretch so far, even if it is true that the sinking of the ships and their subsequent salvage kept Scapa Flow before the public eye for another decade.

Sealyham terrier

Sealyham terriers, short-legged and powerful dogs with massive heads, coloured white or white with brown or lemon markings, were first bred by John Edwardes in the 1840s. Dissatisfied with the sporting dogs available, he decided on a ruthless programme for improvement. Breeding from the toughest terriers available, he then selected only those terriers which proved themselves good killers to produce the next generation. His terriers were called Sealyhams after his estate in Pembrokeshire.

Sedan chair

Sedan chairs take their name from the French fortress town which was the scene of the crushing defeat of Napoleon III's army by German forces on 1 September 1870. Long before then sedan chairs had spread first to Paris, then in 1634 to London. In the eighteenth century they became a marked feature of social life and the 'chairmen' who carried them entered the mythology of the period, rather like the London cabbies who are so often quoted as the embodiment of earthly wisdom. The chair on which the passenger sat was enclosed on three sides by wooden partitions with windows and painted panels, and on the fourth by a door. It was carried through the streets – and even upstairs in great houses – by two chairmen who supported it on poles inserted through fitments on either side. Sedan chairs continued in use for invalids at spas until the First World War.

Shanghai

Around 1870 American sailors started using the verb 'to shanghai' to refer to the practice of drugging a man and putting him aboard a ship to serve in the crew once he had recovered from the inevitable hangover. Shanghai, a seaport on the Hang Pu river in the Yangtze delta, was opened up to international trade under the Treaty of Nanking in 1842, and during the nineteenth century it developed very rapidly. According to Webster's, American sailors were unwilling to sign on for the long trip to China, so unorthodox recruiting methods had to be adopted. Partridge, on the other hand, suggests that American sailors had to be shanghaied to get them to leave a seaport where there were ample opportunities for rest and recreation in congenial if not very choice company. In Australian English, a shanghai is a catapult. Whether it was originally used by or against the native population is uncertain, but it seems unlikely that the Australians would have taken much persuading to retaliate had they been shot at first.

Sheffield plate

The principle of Sheffield plating was discovered in 1743 by one Thomas Bólsover, a cutler active in that city, who devised means of placing a thin coating of silver on sheets of copper. At first he was content to use his process for manufacturing silver buttons, but as he improved the product he went on to use it for all kinds of hollow ware. During the Georgian period, when what was felt to be a heavy tax was levied on silver ware, Sheffield plate became extremely popular with the middle classes.

Sherry

Sherry, like most wines, is called after the place whence it comes, but the name has been shaped more than most by the inability of Englishmen – especially when under the influence – to get their tongues around foreign words. 'Sherry' developed by way of 'Jerries', which perhaps had unfortunate overtones, and then 'Sherries' – which presumably sounded too like a plural to survive – from Jerez, a town in the Spanish province of Andalusia where this fortified wine is produced. Sherry has always tickled English palates, though it is doubtful whether Jack Falstaff would have appreciated a modern chilled Tio Pepe. In Elizabethan times the wine was in fact more generally known as sack, which is a corruption of the adjective 'seco' (or dry) in 'jerez seco', or dry sherry. Williams & Humbert use 'Dry Sack' as a registered trademark, and I don't suppose any pedantry about that really meaning 'Dry Dry' will cause them the least consternation.

Shillelagh

The sound of the word points to the Emerald Isle, and sure enough the cudgels, originally of blackthorn or oak, take their name from a village in County Wicklow, a seaport on the east coast about thirty miles south of Dublin.

Shirley poppies

It is obvious enough that Iceland poppies, which are white or yellow and have the great advantage of lasting well as cut flowers, have their roots in northern regions, and no great imagination is needed to link the colourful Californian poppies (*Eschscholtzia*) with the United States. With that hint, nobody will be too surprised to learn that Shirley poppies are named not after a girl but after a place; what was once the village of Shirley near Croydon. Shirley poppies are a strain of *Papaver rhoeas*, the common red poppy which used to be seen in cornfields and now gives one more splash of brilliant colour to fields of oil-seed rape when the yellow fades away in high summer. Credit for developing the strain around 1880 goes to the Reverend W. Wilks. The flowers, which are generally double, have no trace of black but are found in all shades of white, pink and red.

Shop

Napoleon is supposed to have said that the English were a nation of shopkeepers, and it might be a mistaken quest to trace 'talking shop' back to a place name. All the same this can be done. In Army officers' parlance, the 'Shop' *par excellence* is the Royal Military Academy at Woolwich, and that term is recorded as early as 1841. 'Talking shop' in the mess was, of course, regarded as behaviour unworthy of a gentleman in those times of purchased commissions and general inefficiency. It seems likely that the expression then gradually spread to other walks of life. All this has nothing to do with the term 'a talking shop', applied to a meeting where there is endless debate that leads to no action; still, it does not seem too silly when one recalls the dictum that 'jawing is better than warring'.

Siamese

Thailand has replaced Siam as the name of the country for all practical and political purposes, but we still talk of Siamese twins and Siamese cats, and there is poetry as well as historical correctness in the preference for the old style in *The King and I*. There are records of the birth of twins whose bodies are connected in different ways from various places and periods, but it was Chang and Eng Bunker who became famous as the Siamese twins and the name has stuck. Born on 11 May 1811 in Maklong, they were joined at the chest. They came to England and soon became famous, not least when in April 1843 they married the Misses Sarah and Adelaide Yates, by whom they had respectively ten and twelve children. Chang and Eng died at the age of 62 on 17 January 1872.

It is something of a relief to turn from these unfortunates to the handsome Siamese cats. They were introduced into Britain in the 1870s, and those with seal points on the face, ears, feet and tail were said to come from the Royal Palace in Siam. Certainly they have an aristocratic air.

Solecism

Since Elizabethan times a solecism has been a rather superior word used in condemning any flagrant breach of the rules of grammar or any mispronunciation which betrays ignorance. Within a few years of its introduction into our language it had taken on the transferred sense of an offence against etiquette and good manners, and it is probably in the alliterative expression 'social solecism' that the word is most often employed nowadays. The original solecisms are supposed to have been coined in Soli, an ancient town on the coast of Cilicia, in Asia Minor. Colonists from Argos who settled there allowed their Greek to become so corrupt that it attracted the scorn of their stay-at-home cousins.

Spa

Spas are named after Spa, a town in Belgium about twenty miles south-east of Liège where mineral springs were first discovered in 1326. Much frequented in the Middle Ages and the Renaissance, it became the resort of the rich and famous in the eighteenth century, but suffered a serious setback when devastated by fire in 1807. In English 'spa' has been used since the sixteenth century to mean a place where one can take a cure and enjoy oneself between gulps of evil-tasting waters. Mineral springs were discovered at Leamington as early as 1586, but it was not until 1786 that William Abbotts built what he called the Spa bath; then, in 1838, Queen Victoria granted the designation 'Royal' and the town's cup of satisfaction was brimming over. Her eldest son regularly took a cure at Homburg-von-der-Höhe; it was from there that he brought back the Homburg, a stiff felt hat with a pronounced fore-and-aft dent in the crown which was very popular in the 1890s.

Spaniels

The name 'Spaniel' came into English via the French adjective for Spanish as early as Chaucer's day. The dogs were very popular in England in the sixteenth century, and Shakespeare comments with feeling on their temperamental qualities. Cocker Spaniels were bred to retrieve woodcock.

Stilton

Stilton is a village in Huntingdonshire. Until by-passed by the Great North Road, it was an important staging post for traffic on the way south to London. Stilton is not itself a cheese-making centre, but farmers from Leicestershire used to bring their cheeses to the village in order for them to be carried down to the capital, and our dinner-tables have been suitably enriched.

Stoicism

When we say that Steve Davies 'took it philosophically' when he missed an easy pot, we pay unconscious tribute to Stoicism's enormous influence on European thought. Zeno, who founded Stoicism in the fourth century BC, taught that men should accept the inevitable and irreparable with unruffled composure, nobly displaying their mental superiority over physical forces over which they had no control. Stoicism takes its name from the Stoa Poecile (or Painted Porch) at Athens, where Zeno first expounded the doctrines which proved attractive to men of the calibre of Marcus Aurelius and Seneca and left their mark on the thinking of St Paul.

Strontium

Strontium was named after Strontian, a village on Loch Sunart in Argyllshire, Scotland, where strontium carbonate is found. Discovered in 1808 by Sir Humphry Davy (of safety-lamp fame), strontium is a reactive alkaline-earth metal. People suddenly started taking a serious interest in strontium when it was realized that the isotope Strontium 90, which is a product of nuclear fall-out and emits high-energy beta-rays, can easily become incorporated in bone and cause serious damage to health.

Surrey

Surreys were lightly-built four-wheeled carriages which really did have fringes on top, around the canopy that protected the two passengers from the sun when they went off for a jaunt. The name points to origins in the English county of Surrey, but credit for the design goes to the New York firm of J. B. Brewster & Co. which started producing a version of the English model in 1872. With the invention of the internal combustion engine, the name was applied to horseless carriages of more or less the same pattern. Tilburies – also light carriages, though with only two wheels – were fashionable in the nineteenth century too; but they took their name from their inventor, not from the Essex town by the Thames. Victorias, four-wheeled carriages, were named after the Queen; according to an authority cited by *O.E.D.* the term was coined by the French in 1844, but the modern French lexicographer Robert finds no example of its use in French earlier than 1867.

Sussex pig

Sussex pigs belong not to pork-rearing but to pottery manufacture. A Sussex pig is in fact a pottery jug with a head which can be removed and used as a cup. It appears they were made only at Cadborough, near Rye, in the nineteenth century.

Swedes

Swedes, if not the most exciting vegetable on the table, certainly occupy a place of some significance in the reform of agriculture in the second half of the eighteenth century which helped Britain to feed its increasing industrial population. On retiring from politics in 1730 Charles, second Viscount Townshend, devoted himself to the improvement of his estates around Raynham in Norfolk. One of his most influential innovations was to introduce the cultivation of turnips in fields on a large scale so that they could be fed to cattle throughout the winter months. The Swedish turnip (or Swede, for short) was brought to this country in the 1780s. Swedes are larger than turnips and their white or yellow flesh is coarser, while their hairy leaves grow from a short stem. On Burns' Night, Scots regale themselves to the sound of the pipes on haggis and bashed neeps. This is a reminder of the origin of half the word 'turnip', even if it is uncertain exactly where the prefix 'tur' comes from; and parsnip is a member of the same family of words. Other vegetables with topographical associations include calabrese, Savoy cabbage, Chinese radish and, of course, Brussels sprouts.

Sybarites

Sybarites, who are devoted to the pleasures of the table and the senses, were originally the inhabitants of Sybaris, a Greek colony on the shores of the Gulf of Tarento in southern Italy. Founded by Isus of Helice around 720 BC, the city flourished and by the sixth century had become a byword for luxury and opulence. But high living made the inhabitants soft and in 510 BC the city was attacked and totally destroyed by forces from nearby Croton.

Tangerine

The citron, the ancestor of the orange in its varieties, was first noted as the Median fruit and when Pliny, the Roman writer on natural history, referred to it early in the Christian era, he called it the Assyrian fruit. Tangerines are little oranges which hail from Tangiers, the Moroccan port at the entrance to the Mediterranean; it appears that they were not imported into England until the middle of the nineteenth century.

Tasmanian devils

Tasmanian devils sound pretty terrifying. They are carnivorous marsupials of stocky build, which weigh up to 20 lbs. Their coats are black, sometimes with white spots on throat and rump. For food they will eat any flesh dead or alive, and they are capable of pulling down animals, including lambs, which are bigger than themselves. Their call is a whining growl, following by a snarling yelp. Tasmanian devils are almost entirely confined to the island of Tasmania nowadays, their disappearance from the Australian mainland being due to the dingo. Efforts are being made to save them from extinction by conserving the dense, wet woodlands which form their natural habitat. Brave souls have kept Tasmanian devils as pets, finding them affectionate and clean animals when reared in captivity.

Toc H

Toc H is an acronymic toponym encoded according to British military signallers' practice in the First World War. T H, or 'Toc Aitch' as soldiers were taught to say for clarity when sending messages, stands for Talbot House, a soldiers' club in Poperinghe in the Ypres Salient. It was opened in memory of G. W. L. Talbot who had been killed in 1915. Under the leadership of the Reverend P. T. B. Clayton (1885–1972) Toc H became an important association devoted to Christian fellowship and mutual aid.

Tokay

Tokay is a sweet white wine which has enjoyed a fabulous reputation over the years. It comes from the hilly slopes around the town of Tokaj in north-east Hungary, at the foot of the Hegyalja Mountains. The grapes, which are left on the vines to dry and burst, give a wine which after slow fermentation has a high alcohol content and is rich in natural sugar.

Trappistine

Trappistine is greenish-yellow liqueur made by blending herbs with Armagnac (mentioned under Cognac). It is made according to a recipe kept at the Abbey of Grace Dieu in the Doubs region of France. Like Grand-Marnier, the orange liqueur produced by the firm of Marnier-Lapostolle, Trappistine is much used in cooking.

Trappists

The Trappists, one might imagine, would be unlikely to chatter about the origins of the name by which their Order of Cistercians of the Strict Observance is generally known. In fact it comes from La Trappe, an ancient abbey near Soligny in the diocese of Sées in Normandy, where Armand de Rancé brought in a series of reforms designed to develop spirituality by the exercise of ascetiscism in 1664.

Tridentine

The Tridentine mass has nothing to do with tridents and only indirectly something to do with triple tiaras. It is a version of the Catholic liturgy approved by the great reforming Council of Trent which met at Triento, a city in northern Italy, between 1545 and 1563.

Trojan

To work like a Trojan is, as *O.E.D.* puts it, to work with energy and endurance. This is the triumph of an idiom over history, because all that effort came to nothing when Troy fell after ten years. In earlier times alliteration tempted our ancestors to talk of 'trusty Trojans', but again events cast a cynical light on high principles. The Trojans, with the glaring exception of Paris, were no doubt honourable men and the Greeks were indeed disputatious, vainglorious and devious . . . but their trick gave them victory.

Troy

Troy weight has nothing to do with King Priam's Troy, but is a system of weights for gold and silver inherited from the French city of Troyes on the Seine, about 100 miles south-east of Paris. It was in 1414 that the Troy pound was first mentioned in a statute of Henry V for the regulation of goldsmiths and silversmiths. In Troy weights 24 grains = 1 pennyweight (dwt), 20 dwt = 1 ounce (oz) or 31.1035 grammes; and 12 oz = 1 pound (1 lb) Troy.

In fact the Troy pound went out of use with the passing of an Act of Parliament in 1878 which brought some order out of the chaos of British weights and measures, but Troy ounces have been retained to confuse us all.

Tunbridge ware

Tunbridge ware is very pretty and eminently collectable; it is also hard to describe without making it sound like so much kitsch. Over a period of about two hundred years, the craftsmen of Tunbridge Wells in Kent were famous for their ingenuity in fashioning handy and attractive wooden objects for visitors to the spa. There was a lot of turnery, which was often decorated with marquetry. In the eighteenth century 'Tunbridge ware' became a general term for high-quality turnery and much was quite openly produced under that name in London. From around 1830 until the end of the Victorian era, wooden mosaics were especially popular. The pictures – often of local scenes – and the geometrical patterns were built up painstakingly with thousands of little slips of wood. Great stress was always placed on the fact that the colour of the woods was natural and not produced by dyes, so that there was no risk of fading. In fact there was some cheating to produce the necessary variation in shades, and greys and blues were often obtained by soaking white wood in the chalybeate spring water which had given Tunbridge Wells its original prosperity.

Turquoise

It might appear as obvious that turquoise comes from Turkey – by way of the French 'pierre turquoise' – as that the Cairngorm which graces Highland dress comes from the Scottish mountain.

But there is some scholarly doubt about the matter. Nowadays the word is more commonly used to refer to the bluish-green colour than to the stones – of which the best are said to come from Iran in any case. Also from Turkey come Turkish baths and Turkish delight, the jelly sweetmeat. But the birds we call turkeys have nothing to do with that part of the world; Europeans first came across them in Mexico in 1518. They had aleady been domesticated and introducing them into Europe caused few problem

Tuxedo

Tuxedos were first worn by the smart Americans who frequented the Tuxedo Club which Pierre Lorillard opened on his estate near Lake Tuxedo in New York State in 1886. Though known in England, the word does not appear to be making much progress, even though dinner jacket would hardly seem to be able to offer serious resistance, especially now that the trendy 'DJ' of the fifties has been annexed for a different purpose.

Tweed

Tweed sounds as if it is a toponym, which probably explains why it caught on so readily. In fact it is a corruption of 'tweel', the Scottish form of the word that comes in English as 'twill', and has only comparatively recently become the common name for stout, warm woollen cloth. The manufacture of tweed on the Island of Harris is, however, strictly regulated and in Harris tweed we have a genuine toponym grafted on to a fallacious one.

Ulster

Like the Inverness cape, the Ulster was part of the Victorian gentleman's defence against cold weather. It appeared around 1869 and was distinguished from other varieties of overcoat by being long – reaching down to the ankles sometimes – and having a belt, detachable hood and, a little later, a cape. By 1876 Ulsters were being worn by many, even in the City, but within ten years they had come to be associated more with travelling.

Utopian

Utopian is one of the most succulent of toponyms, because Utopia is a place that never existed. The very word makes that plain: in Greek *U* means no, and *topos* means place. Sir Thomas More published his *Utopia* in Latin in 1516; it contains a description of an imaginary commonwealth where all is governed by reason and gold is despised as worthless. Though many other writers have been tempted to write on Utopias since the Renaissance, the adjective 'utopian' has become a term of abuse in political debate, as being the opposite of practical and sensible. That is a sad fate for a word from a witty and thought-provoking book.

Vaudeville

Vaudeville is a word which has had a remarkably varied life. It seems pretty clear that it all began with one Olivier Basselin, a fifteenth-century fuller who wrote drinking songs which became popular in the valley of the river Vire (i.e. the *vau de Vire*) in Normandy. Next 'vaudeville' came to mean satirical verses, before being used to refer to light comedies interspersed with little songs which were generally performed to familiar airs. In nineteenth-century London, musical comedy was generally called vaudeville, while in the USA the term is applied to what in England used to be called music hall and now passes under the vague designation of variety.

Venetian blinds
As well as Venetian windows – consisting of three parts, of which the outer two are square-headed and the central section is not only wider but arched – and Venetian glass, whose quality was renowned, the Queen of the Adriatic has given us Venetian blinds. Originally made of slats of wood, they were designed to let in the air while excluding prying gazes. It seems likely they were first brought to Venice from somewhere further east.

Waler
Waler was an Indian Army name, first recorded in 1849, for a horse imported to the sub-continent from Australia, especially New South Wales. As well as being a state, New South Wales – like many other places in the New World and the Pacific Basin – is itself a toponym, Captain Cook having given this name to the southern part of the east coast of Australia because it reminded him of South Wales.

Wardour Street
Wardour Street in London, running north-south from Oxford Street to Shaftesbury Avenue, was famous for shops specializing in antique and reproduction furniture. Towards the end of the Victorian era, Wardour Street became a term of condescending condemnation for writing in a pseudo-archaic style in poetry and fiction.

Waterloo
As well as a railway terminus, about which it is perhaps politic to say no more, Wellington's great victory over Napoleon on Sunday 18 June 1815, in the countryside a few miles south of Brussels, has given us the phrase 'to meet one's Waterloo'. There

is nothing quite the same in French, although there is an expression 'un coup de Trafalgar'; its meaning, however, is 'a dastardly trick', which properly ought to refer to the way Nelson was shot by marksmen stationed in fighting tops of the French man-of-war *Redoubtable*, rather than to English tactics in the battle.

As well as marking the end of Napoleon's ambitions, Waterloo may be the beginning of 'loo', this generation's genteelism for what our Elizabethan forefathers called the jakes. The theory is that someone saw a kind of pun between Waterloo and water closet. However, there are other suggestions. One is that it comes from 'gardyloo', which is what Edinburgh people yelled out before flinging the contents of their chamber-pots out of their bedroom windows into the street below; the actual cry – and possibly the repulsive practice – is supposed to have been picked up from French aristocrats in the Scottish capital in the Renaissance. Other ingenious philologists argue that 'loo' is a mistaken attempt to pronounce the number '100' which used to be painted on lavatory doors in French hotels.

All this has virtually nothing to do with the card game called loo which was popular in the eighteenth century. That name comes from 'lanterlu', the nonsense refrain of a seventeenth-century French comic song. Irish loo is defined with delicious economy as five-card loo played with three cards.

Welsh Rabbit

Dictionaries seem unanimous in the view that it is a mistake to imagine 'Welsh rarebit' to be the correct form; they explain that as a popular etymology, an unscientific attempt to make sense of the name. However that may be, it is odd to think that with

Irish stew and haggis, it stands as a nation's major culinary export. A recipe that is supposed to be traditional runs:

8 oz strong cheese, grated or cut small	1 teaspoon mustard
1 tablespoon butter	2 teaspoons flour
liberal dash Worcestershire sauce	4 tablespoons beer (or milk)
	pepper

Heat gently in a saucepan until a thickish paste. Spread on slices of bread toasted on one side only and brown the top under the grill.

Welsh rabbit makes a grand savoury for the end of a banquet and sets one up for enjoying the port afterwards. This is more than can be said for Buck rabbit, which is Welsh rabbit with a poached egg on top.

Winchester

Winchester repeating rifles were manufactured by Oliver Fisher Winchester (1810–1880) at his works in New Haven, Connecticut. Winchester quart bottles, however, which may still be found in laboratories and pharmacies, have a genuine connection with the old city in Hampshire. Containing 80 oz (or half a gallon), they are the lone survivors of an 1835 Act of Parliament which abolished a whole series of so-called Winchester measures.

Worsted

Of all the countless names for types of cloth derived from the places where they originated, none is more famous than worsted and it seems perfectly possible that the spelling is as authentic as the present-day form of Worstead. That pleasant Norfolk village with its beautiful church was once a centre of the East Anglia wool trade, specializing in tightly twisted long-staple yarn which made up into fine cloth. The term has been in use since the mid-fifteenth century, though the trade left Worstead long since.

Yorkshire

As well as Yorkshire pudding, the proof of which is in the eating and not the etymology, the county has given us Yorkshire Fog. This is not a meteorological condition but a plant, in Latin *Holcus lanatus*, which is known in North America as Velvet Grass. Regarded in most situations as no better than a weed, Yorkshire Fog has some value as grazing on poor lands, such as moors, before it becomes too old and unpalatable. In campanology Yorkshire Surprise is one of the many standard methods in change ringing named after counties and large cities; there is also Pudsey Surprise. For a Surprise Major eight bells are needed, for Royal there must be ten and Maximus requires no fewer than twelve. When a bowman shoots a York round in an archery competition he shoots 72 arrows at 100 yards, 48 at 80 yards and 24 at 60 yards. In cricket a yorker is a fast or at least medium-paced ball which pitches at exactly the place where the batsmen rests his bat as he awaits the delivery. It comes to him quickly, leaving him only a split second to decide whether to advance bravely and turn it into a full toss, or to go back and play it defensively. As Joel Garner shows, it is very effective against batsmen who have just come in, especially those in the lower half of the order, and generally makes for better cricket than banging in bouncers to whistle around their ears.

The term was coined around 1870 and it seems probable that it was a Yorkshire speciality.

Ytterbium

Ytterbium would probably not rate an entry here if it did not occupy a rather distinguished place in any alphabetical sequence of toponyms. It is a chemical element in the rare earth group and takes its name from Ytterby, a town in Sweden not far from Stockholm. The oxide ytterbite was first recorded in 1839, and the element ytterbium was separated out in 1878. That was not the end of the story, however, for it was further investigated by C. Matignon in 1907 , when he concluded that there were in reality two rare earths here, one of which he called Neo-Ytterbium while naming the other Lutecium, i.e. the Paris element. It seems that chemists have found little use for these rare earths, which is a pity considering how valuable ytterbium is to hard-pressed lexicographers.

Zionism

Zionism is perhaps most simply summed up in the words of the Balfour Declaration of 2 November 1917; it is the policy that seeks 'the establishment in Palestine of a National Home for the Jewish People'. When in exile in Babylon in ancient times the Jews had longed to return to Jerusalem, and their prayers had been answered. In more modern times Jews who were dispersed in many lands where they suffered from persecution and discrimination, dreamt of a better life in Palestine. In 1896 Theodor Herzl did much to focus thought on the issue when he published his book called *The Jewish State*, and a year later the First Zionist Conference meeting in Basle defined the aims of Zionism. Developments during two World Wars are too complex to trace here, but the outcome was the proclamation of the state of Israel by the redoubtable David Ben Gurion on 14 May 1948. Since then the stormy history of Israel's struggle for survival has filled the headlines of the world's newspapers.

A coinage dating from 1896, Zionism comes of course from what Bible-readers and hymn-singers know better as Sion, one of the hills on which Jerusalem stands. Yet another variant spelling is found in Syon House, the Duke of Northumberland's great house in Isleworth, which stands on the site of the Brigittine Syon convent founded by Henry V in the year of Agincourt and moved out to Isleworth in 1431.

Zwieselite

With deadpan impenetrability, *O.E.D.* defines Zwieselite as a clove-brown variety of triplite, which in its turn is described as a phosphate of iron and manganese, with cleavage in three directions mutually at right-angles. Zwieselite takes its name from Zwiesel, a town in Bavaria, and was first identified in 1861.